Collins

# Weather
## ALMANAC

### A GUIDE TO
## 2023

## Storm Dunlop

Published by Collins
An imprint of HarperCollins Publishers
Westerhill Road
Bishopbriggs
Glasgow G64 2QT
www.harpercollins.co.uk

HarperCollins*Publishers*
Macken House, 39/40 Mayor Street Upper,
Dublin1, D01 C9W8, Ireland

A catalogue record for this book is available from the British Library

ISBN 978-0-00-853260-4

10 9 8 7 6 5 4 3 2

Printed in the UK using 100% Renewable Electricity at
CPI Group (UK) Ltd

If you would like to comment on any aspect of this book,
please contact us at the above address or online.
e-mail: collins.reference@harpercollins.co.uk

This book is produced from independently certified FSC™ paper
to ensure responsible forest management.

For more information visit: www.harpercollins.co.uk/green

# Contents

# Introduction

**Our variable weather**

Anyone living in the British Isles hardly needs to be told that the weather is extremely variable. Despite what politicians and the media may say, the extreme weather sometimes experienced in Britain is not 'unprecedented'. Britain has always experienced extreme windstorms, snowfall, rainfall, thunderstorms, flooding and such major events. And always will. Such events may be unusual, and 'not within living memory of the oldest inhabitant', but, overall, the weather always exhibits such extremes. We may have learned from events such as the East Coast floods of 1953, and constructed the Thames Barrier and the barrier on the River Hull, but a North Sea surge will occur again. The Somerset Levels have been flooded many times in the past, so the flooding in 2012 and 2014 was not that extraordinary. They were flooded in 1607, so have a long history of flooding. Global warming may produce more occasions when extreme events occur, but no meteorologists can predict when these may happen.

The weather in Britain – its climate – is basically a maritime climate, determined by the proximity of these islands to the Atlantic Ocean. It is largely determined by the changes resulting from incursions of dry continental air from the Eurasian landmass to the east, contrasting with the prevailing moist maritime air from the Atlantic Ocean to the west. The general mildness of the climate, in comparison with other locations at a similar latitude, has often been ascribed to 'the Gulf Stream'. In fact, the Gulf Stream exists only on the western side of the Atlantic, along the East Coast of America, and the warm current off the coast of Britain is correctly known as the North Atlantic Drift. In reality, it is not solely the warmth of the oceanic waters that creates the mild climate.

The Rocky Mountains in North America impede the westerly flow of air and create a series of north/south waves that propagate eastwards (and actually right round the world). These waves cause north/south oscillations of the jet stream (which, in our case, is the 'Polar Front' jet stream) that, in turn, controls the progression of the low-pressure areas (the depressions) that travel from west to east and create most of the changeable weather over Britain. The jet stream 'steers' the depressions, sometimes to the north of the British Isles and sometimes far to the south. It may also assume a strong flow in longitude (known as a 'meridional' flow) leading to a blocking situation, where depressions cannot move eastwards and may come to a halt or be forced to travel far to the north or south. Such a block (especially in winter) may draw frigid air directly from the Arctic Ocean or continental air from the east (such air is usually described as 'from Siberia', because it often originates from that region). More rarely, such blocks draw warmer air from the Mediterranean over the country.

The location of the jet stream itself is governed by something known to meteorologists as the North Atlantic Oscillation (NAO). Basically, this may be thought of as the distribution of pressure between the Azores High over the central Atlantic and the semi-permanent Icelandic Low. These are semi-permanent features of the flow of air around the globe, and are known to meteorologists as 'centres of action'. The average route of the jet stream is to the north of the British Isles, so the whole country is subject to the strong westerlies that prevail at these latitudes. Northern regions of the British Isles tend to be mainly subject to the depressions arriving from the Atlantic and the weather that accompanies them. So overall, the climate of the British Isles may be described as: warmer and drier in the south and east, wetter in the west and north.

When the NAO has a 'positive' index, with high pressure and warm air in the south compared with low pressure and low temperatures in the north, depressions are steered north of the British Isles. Although most of the country experiences windy and wet weather, the west and north tends to be affected most, with southern and eastern England warmer and drier. When the NAO has a 'negative' index, the Azores High is displaced towards the north and the jet stream tends to show strong meanders towards the south. Frequently there is a slowly moving low pressure area over the near (north-western) Continent or over the north-eastern Atlantic. The jet stream wraps around this, before turning back to the north. Blocks are, however, most frequent in the spring.

**Climatic regions of the British Isles**
It is appropriate to deal with the climates of the British Isles by discussing eight separate regions, which are:

1 South West England and the Channel Islands (page 214)

2 South East England and East Anglia (page 216)

3 The Midlands (page 218)

4 North West England and the Isle of Man (page 220)

5 North East England and Yorkshire (page 222)

6 Wales (page 224)

7 Ireland (page 226)

8 Scotland (page 229)

These different regions are discussed later in this book.

**Sunrise / Sunset & Moonrise / Moonset**
For four dates within each month, tables show the times at which the Sun and Moon rise and set at the four capital cities in the United Kingdom: Belfast, Cardiff, Edinburgh and London. For Edinburgh and London, the locations of specific observatories are used, but not for Belfast and Cardiff, where more general locations are employed. Although the exact time of Sunrise/Sunset and Moonrise/Moonset at any observer's location depends on their exact position, including their latitude and longitude and height above sea level, the times shown will give a useful indication of the timing of the events. However, calculation of rising and setting times is complicated and strongly depends on the location of the observer. Note that all the times in this book are calculated astronomically, using what is known as Universal Time (UT), sometimes known as Coordinated Universal Time (UTC). This is identical to Greenwich Mean Time (GMT). The times given do not take account of British Summer Time (BST). Some effects are quite large. For example, at the summer solstice in 2023, sunrise is some 47 minutes later at Edinburgh than it is at Lerwick in the Shetlands. Sunset is just 31 minutes earlier on the same date.

**Edinburgh & Lerwick: Comparison of sunrise & sunset times**
A comparison of the timings of sunrise and sunset at Lerwick (latitude 60.16 N) and Edinburgh (latitude 55.9 N) in the following table gives an idea of how the times change with latitude. The timings are those at the equinoxes (March 20 and September 23 in 2023) and the solstices (June 21 and December 22 in 2023). It is notable, of course, that, although the times are different, the azimuth of sunrise is identical at both equinoxes, because the Sun is then crossing the celestial equator. At the solstices, both times and azimuths are different between the two locations. Azimuth is measured in degrees from north, through east, south and west, and then back to north.

| Location | Date | Rise | Azimuth | Set | Azimuth |
|----------|------|------|---------|-----|---------|
| Edinburgh | 20 Mar 2023 (Mon) | 06:16 | 89 | 18:26 | 271 |
| Lerwick | 20 Mar 2023 (Mon) | 06:07 | 89 | 18:18 | 271 |
| Edinburgh | 21 Jun 2023 (Wed) | 03:26 | 43 | 21:03 | 317 |
| Lerwick | 21 Jun 2023 (Wed) | 02:39 | 34 | 21:34 | 326 |
| Edinburgh | 23 Sep 2023 (Sat) | 05:59 | 89 | 18:10 | 271 |
| Lerwick | 23 Sep 2023 (Sat) | 05:50 | 89 | 18:02 | 271 |
| Edinburgh | 22 Dec 2023 (Fri) | 08:42 | 134 | 15:40 | 226 |
| Lerwick | 22 Dec 2023 (Fri) | 09:08 | 141 | 14:58 | 219 |

Apart from the times, this table (and the monthly tables) also shows the azimuth of each event, which indicates where the body concerned rises or sets. These azimuths are given in degrees, and the table given here shows the azimuths for various compass points in the eastern and western sectors of the horizon.

**Table of azimuths**

| Degrees | Compass point |
| --- | --- |
| *Eastern horizon* | |
| 45° | NE |
| 67° 30' | ENE |
| 90° | E |
| 112° 30' | ESE |
| 135° | SE |
| 157° 30' | SSE |
| *Western horizon* | |
| 202° 30' | SSW |
| 225° | SW |
| 247° 30' | WSW |
| 270° | W |
| 292° 30' | WNW |
| 315° | NW |
| 337° 30' | NNW |

The actual latitude and longitude used in the calculations are shown in the following table. It will be seen that the altitudes of the observatories in Edinburgh (Royal Observatory Edinburgh, ROE) and London (Mill Hill Observatory) are quite considerable (that for ROE is particularly large) and these altitudes will affect the rising and setting times, which are calculated to apply to observers closer to sea level. (Generally, close to the observatories, such rising times will be slightly later, and setting times slightly earlier than those shown.)

**Latitude and longitude of UK capital cities**

| City | Longitude | Latitude | Altitude |
| --- | --- | --- | --- |
| Belfast | 5°56'00.0" W | 54°36'00.0" N | 3 m |
| Cardiff | 3°11'00.0" W | 51°30'00.0" N | 3 m |
| Edinburgh (ROE) | 3°11'00.0" W | 55°55'30.0" N | 146 m |
| London (Mill Hill) | 0°14'24.0" W | 51°36'48.0" N | 81 m |

### Twilight

For each individual month, we give details of sunrise and
sunset times (and Moonrise and Moonset times), together with
the azimuths (which give an idea of where the rising or setting
takes place – see previous page) for the four capital cities of
the regions of the United Kingdom. But twilight also varies
considerably from place to place, so the monthly diagrams
here show the duration of twilight at those four cities. During
the summer, twilight may persist throughout the night. This
applies everywhere in the United Kingdom, so two additional
yearly twilight diagrams are included (on pages 259 and 261):
one for Lerwick in the Shetlands and one for St Mary's in the
Scilly Isles. Although the hours of complete darkness increase
as one moves towards the equator, it will be seen that there is
full darkness nowhere in the British Isles at midsummer.

There are three recognised stages of twilight: *civil twilight*,
when the centre of the Sun is less than 6° below the horizon;
*nautical twilight*, when the Sun is between 6° and 12° below the
horizon; and *astronomical twilight*, when the Sun is between 12°
and 18° below the horizon. Full darkness occurs only when the
Sun is more than 18° below the horizon. The time at which civil
twilight begins is sometimes known in the UK as 'lighting-up
time'. Stars are generally invisible during civil twilight. During
nautical twilight, the very brightest stars only are visible. (These
are the stars that were used for navigation, hence the name
for this stage.) During astronomical twilight, the faintest stars
visible to the naked eye may be seen directly overhead, but are
lost at lower altitudes. They become visible only once it is fully
dark. The diagrams show the duration of twilight at the various

cities. Of the locations shown, during the summer months there is astronomical twilight at most of the locations, except at Lerwick, but there is never full darkness during the summer anywhere in the British Isles.

The diagrams also show the times of New and Full Moon (black and white symbols, respectively). As may be seen, at most locations during the year, roughly half of New and Full Moon phases may come during daylight. For this reason, the exact phase may be invisible at one location, but be clearly seen elsewhere in the world. The exact times of the events are given in the diagrams for each individual month.

Twilight diagrams for each of the four capital cities are shown every month, and full yearly diagrams are shown on pages 259–261.

Also shown each month is the phase of the Moon for every day, together with the age of the Moon, which is counted from New Moon.

**The seasons**

By convention, the year has always been divided into four seasons: spring, summer, autumn and winter. In the late eighteenth century, an early German meteorological society, the Societas Meteorologica Palatina, active in the Rhineland, defined the seasons as each consisting of three whole months,

---

*Doldrums*

The doldrums are a zone of reduced winds, generally located over the equatorial region, although moving north and south with the seasons. Air in the Doldrums is largely rising, because of solar heating, and horizontal motion across the surface is reduced or non-existent. The Doldrums were thus a major obstacle for sailing ships and to become becalmed in the Doldrums was a significant hazard to those undertaking long ocean voyages.

---

beginning before the equinoxes and solstices. So spring consisted of the months of March, April and May; summer of June, July and August; autumn of September, October and November, and winter of December, January and February. There has been a tendency by meteorologists to follow this convention to this day, with winter regarded as the three calendar months with the lowest temperatures in the northern hemisphere (December, January and February) and summer those with the warmest (June, July and August). Astronomers, however, regard the seasons as lasting three months, but centred on the dates of the equinoxes and solstices (March 20, September 23 and June 21, December 22 in 2023).

Some ecologists tend to regard the year as divided into six seasons. Analysis of the prevailing weather types in Britain, however, suggests that there are five distinct seasons. (The characteristic weather of each is described in the month in which the season begins.) Although, obviously, the seasons cannot be specified as starting and ending on specific calendar dates, it is useful to identify them in this way. So in Britain, we have:

**Early winter**
November 20 to January 19 (see page 181)

**Late winter and early spring**
January 20 to March 31 (page 17)

**Spring and early summer**
April 1 to June 17 (page 65)

**High summer**
June 18 to September 9 (page 97)

**Autumn**
September 10 to November 19 (page 148)

# The Weather in 2021

There was heavy snowfall early in January and February over Scotland and Northern Ireland in particular. A snow depth of 32 cm was noted at Trassey Slievenaman in County Down on January 25. The lowest temperature of the year, -23°C was recorded at Braemar (Aberdeenshire) on February 11.

The first named storm was Storm Christoph, which brought wet and windy weather to Wales and Scotland, and heavy snow to parts of Scotland between January 19 and 21. Scotland, Northern Ireland and Wales experienced more snow and ice at the end of January. Bitterly cold easterly winds associated with Storm Darcy brought snow to eastern areas of England, including East Anglia, in early February.

The hottest period came in the middle of July, when several heat health warnings were issued for various parts of the country because of expected temperatures above 30°C. A high of 32.2°C (the highest of the year) was recorded at Heathrow (Greater London) on July 20. Northern Ireland experienced a run of very high temperatures, with 31.3°C recorded at Castlederg in County Tyrone on 21 July. This is an all-time record for Northern Ireland.

Storm Evert was a late-season storm that brought high winds and rain to the south-west of England on July 30. There was an amber warning for Cornwall and the Isles of Scilly, with a yellow warning for the south coast and South Wales.

There was persistent heavy rain in southern Scotland and northern England in October. As usual, Honister Pass in Cumbria saw heavy rainfall, with 222.6 mm of rain in the 24 hours to 9:00 am on 28 October, and this was the maximum for the year.

Storm Arwen arrived in November and the Met Office issued a rare red warning of high winds for the eastern coasts of southern Scotland and northern England. The effects of these winds were felt on November 26 and 27, with many trees uprooted in southern Scotland and major power outages. There was also some disruption in Northern Ireland and as far south as Wales, with both road and rail traffic affected.

The final named storm of 2021, Storm Barra, arrived on December 7. It caused disruption in Northern Ireland with high winds and flooding. It also brought snow to northern England. Wales and parts of southern England, particularly East Anglia, also experienced road and rail disruption mainly because of the high winds.

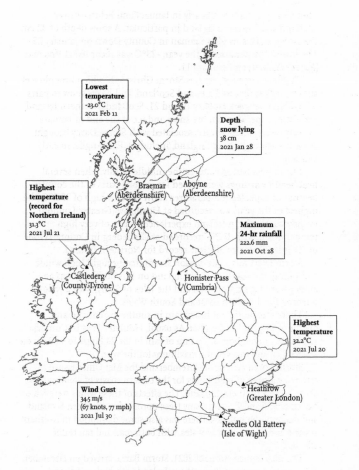

Lowest
temperature
-23.0°C
2021 Feb 11

Depth
snow lying
38 cm
2021 Jan 28

Highest
temperature
(record for
Northern Ireland)
31.3°C
2021 Jul 21

Maximum
24-hr rainfall
222.6 mm
2021 Oct 28

Braemar
(Aberdeenshire)

Aboyne
(Aberdeenshire)

Castlederg
(County Tyrone)

Honister Pass
(Cumbria)

Highest
temperature
32.2°C
2021 Jul 20

Heathrow
(Greater London)

Wind Gust
34.5 m/s
(67 knots, 77 mph)
2021 Jul 30

Needles Old Battery
(Isle of Wight)

January

# Introduction

January sees some of the coldest temperatures of the year. Braemar, the village in the Scottish highlands, about 93 km west of Aberdeen, has twice recorded the lowest temperature in the British Isles (-27.2°C), first on 11 February 1895 and since then on 10 January 1982. Only Altnaharra in Sutherland has ever recorded a similar temperature (on 30 December 1995). Generally, Scotland sees extensive snow cover in January, particularly important at the ski centres at Cairngorm Mountain, Glencoe Mountain, Glenshee, Nevis Range and The Lecht, although global warming threatens to decrease the coverage of snow. Deep snow has become less frequent and the centres have had to invest in snow-making cannons.

The weather in January is often dominated by cold easterly winds. These arise because of the cold anticyclone over the near continent that builds up during most winters. This is often an extension of the great wintertime Siberian High that dominates the weather over Asia and creates a cold, dry airflow over the eastern side of Asia. Circulation round this anticyclone tends to bring a cold airflow across the North Sea. In crossing the sea, the air gains moisture and this often results in snowfall along the eastern coast of Britain. At times, a blocking situation may arise with a major high-pressure area over Scandinavia. This brings extremely cold Arctic air down over the British Isles and, depending on how long the block persists, may give rise to a persistent spell of low temperatures.

Snowfall is, however, very frequent farther south, and the exceptional snowfall in 1947, which was undoubtedly the winter that saw the greatest fall of snow over Britain, did not begin until quite late in the month of January. Although there had been some snow earlier, in December and early January, it had melted by the middle of the month, and there were unseasonably high temperatures across the country. The temperature then dropped and there were frosts at night from January 20 (the nominal beginning of the late winter, early spring season). Snow-bearing clouds began to move into the south-west of England on January 22. There was heavy snowfall with blizzard conditions in the West Country. Even the Scilly Isles saw a slight covering of snow, amounting to a few centimetres in depth – an almost unprecedented event. The following days saw heavy snowfall extend right across all of England and Wales before spreading into Scotland. There were seemingly relentless snowstorms over the next few weeks, which left England and Wales, up as far as the Scottish Borders, buried beneath a deep blanket of snow, and movement by rail and road almost completely paralysed. Somewhat ironically, although snow fell somewhere in the United Kingdom on every day from January 22 until the middle of March, Scotland escaped the worst storms. After some three months of northerly and easterly winds, by March 10, warm southerly and south-westerly winds started to affect the West Country. Not only did the warm airstream bring dense fogs, it also produced heavy rainfall, which in turn led to floods as the rain ran off the frozen ground. The situation was worsened by the gradual thawing of the immense snowpack, leading to even more extreme flooding. By March 13, even the rivers in East Anglia were about to burst their banks.

Only the winter of 1962–63 saw a longer period during which snow persisted, and much lower temperatures, but the amount of snow that fell in 1947 was the most extreme ever recorded for the United Kingdom.

### Late winter and early spring season – January 20 to March 31

This season tends to exhibit long spells of settled conditions. These may be of very cold weather, characterised by Arctic air, introduced by a northerly airflow, and thus forming the main period of winter. Although in some years the weather may take on the character of an extended spring, such conditions are less frequent than those with low temperatures. There may be long spells of wet, westerly conditions, but these tend to be less common than spells with cold northerly or easterly winds. However, the wet westerlies suddenly become less frequent after about March 9, and indeed westerly weather then becomes very uncommon.

# Weather Extremes

| Country | Temp. | Location | Date |
|---|---|---|---|
| **Maximum temperature** | | | |
| England | 17.6°C | Eynsford (Kent) | 27 Jan. 2003 |
| Northern Ireland | 16.4°C | Knockarevan (Co. Fermanagh) | 26 Jan. 2003 |
| Scotland | 18.3°C | Aboyne (Aberdeenshire) Inchmarlo (Kincardineshire) | 26 Jan. 2003 |
| Wales | 18.3°C | Aber (Gwynedd) | 10 Jan. 1971 27 Jan. 1958 |
| **Minimum temperature** | | | |
| England | -26.1°C | Newport (Shropshire) | 10 Jan. 1982 |
| Northern Ireland | -17.5°C | Magherally (Co. Down) | 1 Jan. 1979 |
| Scotland | -27.2°C | Braemar (Aberdeenshire) | 10 Jan. 1982 |
| Wales | -23.3°C | Rhayader (Powys) | 21 Jan. 1940 |

| Country | Pressure | Location | Date |
|---|---|---|---|
| **Maximum pressure** | | | |
| Scotland | 1053.6 hPa | Aberdeen Observatory | 31 Jan. 1902 |
| **Minimum pressure** | | | |
| Scotland | 925.6 hPa | Ochtertyre (Perthshire) | 26 Jan. 1884 |

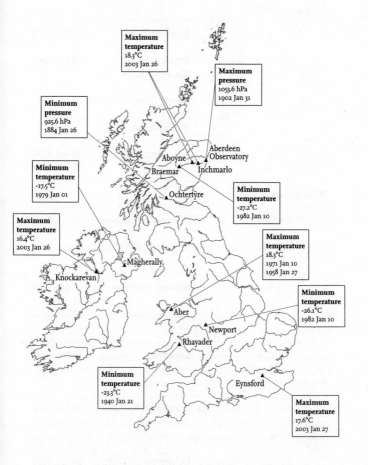

**Maximum temperature**
18.3°C
2003 Jan 26

**Maximum pressure**
1053.6 hPa
1902 Jan 31

**Minimum pressure**
925.6 hPa
1884 Jan 26

**Minimum temperature**
-17.5°C
1979 Jan 01

**Maximum temperature**
16.4°C
2003 Jan 26

**Minimum temperature**
-27.2°C
1982 Jan 10

**Maximum temperature**
18.3°C
1971 Jan 10
1958 Jan 27

**Minimum temperature**
-26.1°C
1982 Jan 10

**Minimum temperature**
-23.3°C
1940 Jan 21

**Maximum temperature**
17.6°C
2003 Jan 27

Aberdeen Observatory
Aboyne
Braemar
Inchmarlo
Ochtertyre
Magherally
Knockarevan
Aber
Newport
Rhayader
Eynsford

# The Weather in January 2022

| Observation | Location | Date |
| --- | --- | --- |
| **Max. temperature** 16.3°C | St James's Park (Greater London) | 1 January |
| **Overnight minimum** 13.2°C | Chivenor (Devon) | 1 January |
| **Min. temperature** -8.0°C | Topcliffe (N. Yorks) | 6 January |
| **Rainfall** 64.8 mm | Honister Pass (Cumbria) | 8 January |
| **Wind gust** 42 m/s (81 knots or 93 mph) | Brizlee Wood (Northumberland) | 29 January |
| **Snow depth** 11 cm | Loch Glascarnoch (Ross & Cromarty) | 5, 6 & 8 January |

The month began with very mild weather. A depression centred to the northwest of Scotland (halfway to Iceland) brought a stream of exceptionally mild south-westerly air to the whole country. It was the warmest January 1 (New Year's Day) for many years. This produced an extremely high temperature of 16.3°C at St James's Park in Central London. This high (still provisional) exceeded the previous highest temperature for that day of 15.6°C, recorded at Bude in Cornwall as long ago as 1916. Many places in the country experienced extremely mild temperatures, and the recorded 13.2°C at Chivenor in Devon on January 1 is also provisionally accepted as a record minimum overnight temperature.

By January 4, the mild air had been replaced by a cold northerly flow originating on the western side of a depression off northern Norway. Temperatures dropped dramatically. There were heavy snow showers in Scotland and a depth of snow of 11 cm was reported on January 5, 6 and 8 at Loch Glascarnoch in Ross & Cromarty. The cold air extended south. There were sharp frosts in Scotland and heavy snow in northern England with many road closures. By January 6, Topcliffe in North Yorkshire recorded the month's minimum of -8°C. Frontal systems then brought heavy rain to certain areas and Honister Pass in Cumbria recorded 64.8 mm in the 24 hours to January 8. Bands of rain crossed the northern region of England in the next few days, being replaced by widespread showers.

A period of relatively settled weather resulted, with widespread fog over most of the country, although with some patchy rain in Scotland, particularly in the north-west. An anticyclone became established over southern Britain by January 13 and this resulted in many southern areas having abundant sunshine, but also patchy fog. High pressure was dominant over practically the whole country for some days. However, the end of the month brought extremely windy weather, with two named storms affecting the country.

The first, Storm Malik, gave extreme winds over northern England and Scotland on January 29, and Brizlee Wood in Northumberland reported gusts to 42 m/s (81 knots, 93 mph) that day. There was widespread disruption in northern England and southern Scotland to travel and power outages. Northern Ireland also experienced rail and road closures. The second named storm, Storm Corrie, brought high winds and driving rain to England, reaching the north by dusk. Wales experienced windy weather and local heavy rain with more general rain spreading southwards on January 31. Storm Corrie was particularly severe over Scotland, with Stornoway in the Western Isles reporting gusts to 41 m/s (about 80 knots or 92 mph).

Overall, it was extremely dry in the south-east and the amount of sunshine meant that it is provisionally the sunniest January for a century.

# Sunrise and Sunset 2023

| Location | Date | Rise | Azimuth ° | Set | Azimuth ° |
|---|---|---|---|---|---|
| **Belfast** | | | | | |
| | Jan 01 (Sun) | 08:46 | 131 | 16:08 | 229 |
| | Jan 11 (Wed) | 08:41 | 128 | 16:22 | 232 |
| | Jan 21 (Sat) | 08:31 | 125 | 16:40 | 236 |
| | Jan 31 (Tue) | 08:15 | 120 | 17:00 | 240 |
| **Cardiff** | | | | | |
| | Jan 01 (Sun) | 08:18 | 128 | 16:14 | 232 |
| | Jan 11 (Wed) | 08:15 | 125 | 16:27 | 235 |
| | Jan 21 (Sat) | 08:06 | 122 | 16:42 | 238 |
| | Jan 31 (Tue) | 07:53 | 118 | 17:00 | 243 |
| **Edinburgh** | | | | | |
| | Jan 01 (Sun) | 08:43 | 133 | 15:49 | 228 |
| | Jan 11 (Wed) | 08:38 | 130 | 16:04 | 230 |
| | Jan 21 (Sat) | 08:26 | 126 | 16:22 | 234 |
| | Jan 31 (Tue) | 08:10 | 121 | 16:43 | 239 |
| **London** | | | | | |
| | Jan 01 (Sun) | 08:07 | 128 | 16:02 | 232 |
| | Jan 11 (Wed) | 08:03 | 125 | 16:14 | 235 |
| | Jan 21 (Sat) | 07:55 | 122 | 16:30 | 238 |
| | Jan 31 (Tue) | 07:42 | 118 | 16:48 | 243 |

*Note that all times are in Universal Time (UT), otherwise known as Greenwich Mean Time (GMT). These times do not take Summer Time (BST) into account.*

# Moonrise and Moonset 2023

| Location | Date | Rise | Azimuth ° | Set | Azimuth ° |
|----------|------|------|-----------|-----|-----------|
| **Belfast** | | | | | |
| | Jan 01 (Sun) | 12:46 | 64 | 03:08 | 292 |
| | Jan 11 (Wed) | 21:15 | 73 | 11:02 | 291 |
| | Jan 21 (Sat) | 09:05 | 140 | 15:37 | 222 |
| | Jan 31 (Tue) | 11:44 | 43 | 04:56 | 316 |
| **Cardiff** | | | | | |
| | Jan 01 (Sun) | 12:44 | 66 | 02:48 | 290 |
| | Jan 11 (Wed) | 21:10 | 74 | 10:43 | 289 |
| | Jan 21 (Sat) | 08:30 | 135 | 15:48 | 226 |
| | Jan 31 (Tue) | 11:54 | 47 | 04:24 | 312 |
| **Edinburgh** | | | | | |
| | Jan 01 (Sun) | 12:30 | 63 | 03:00 | 293 |
| | Jan 11 (Wed) | 21:01 | 72 | 10:54 | 292 |
| | Jan 21 (Sat) | 09:05 | 142 | 15:14 | 220 |
| | Jan 31 (Tue) | 11:22 | 41 | 04:55 | 318 |
| **London** | | | | | |
| | Jan 01 (Sun) | 12:32 | 66 | 02:36 | 290 |
| | Jan 11 (Wed) | 20:57 | 74 | 10:31 | 289 |
| | Jan 21 (Sat) | 08:19 | 135 | 15:15 | 226 |
| | Jan 31 (Tue) | 11:41 | 47 | 04:12 | 312 |

*Note that all times are in Universal Time (UT), otherwise known as Greenwich Mean Time (GMT). These times do not take Summer Time (BST) into account.*

# Twilight Diagrams 2023

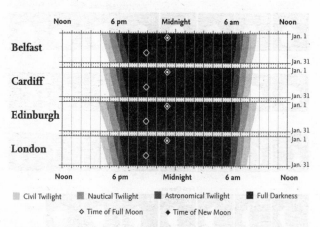

| | Noon | 6 pm | Midnight | 6 am | Noon |
|---|---|---|---|---|---|

Belfast

Cardiff

Edinburgh

London

▨ Civil Twilight    ▨ Nautical Twilight    ■ Astronomical Twilight    ■ Full Darkness

◇ Time of Full Moon    ◆ Time of New Moon

**The exact times of the Moon's major phases are shown on the diagrams opposite.**

## Jet streams

Jet streams are narrow ribbons of fast moving air, typically hundreds of kilometres wide and a few kilometres in depth. The most important one for British weather is the Polar Front Jet Stream, a westerly wind that flows right round the Earth. It is driven by the great temperature difference between the cold polar air and warmer air closer to the equator. Fluctuations in latitude are primarily caused by the flow across the Rockies in North America. These fluctuations in latitude, known as Rossby waves, spread right across the continental United States and across the Atlantic – and even farther. The jet stream has a great effect on the strength of depressions and also on their paths. It may cause depressions to sometimes pass directly across the British Isles and sometimes to the north or south of them.

# The Moon's Phases and Ages 2023

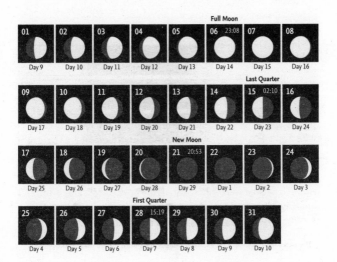

| | | | | | Full Moon | | |
|---|---|---|---|---|---|---|---|
| 01 | 02 | 03 | 04 | 05 | 06 23:08 | 07 | 08 |
| Day 9 | Day 10 | Day 11 | Day 12 | Day 13 | Day 14 | Day 15 | Day 16 |

| | | | | | | Last Quarter | |
|---|---|---|---|---|---|---|---|
| 09 | 10 | 11 | 12 | 13 | 14 | 15 02:10 | 16 |
| Day 17 | Day 18 | Day 19 | Day 20 | Day 21 | Day 22 | Day 23 | Day 24 |

| | | | New Moon | | | | |
|---|---|---|---|---|---|---|---|
| 17 | 18 | 19 | 20 | 21 20:53 | 22 | 23 | 24 |
| Day 25 | Day 26 | Day 27 | Day 28 | Day 29 | Day 1 | Day 2 | Day 3 |

| First Quarter | | | | | | |
|---|---|---|---|---|---|---|
| 25 | 26 | 27 | 28 15:19 | 29 | 30 | 31 |
| Day 4 | Day 5 | Day 6 | Day 7 | Day 8 | Day 9 | Day 10 |

---

*Gulf Stream*
A warm-water current on the western side of the North Atlantic Ocean. It extends along the eastern seaboard of the United States from the Gulf of Mexico to Cape Hatteras. It then turns eastwards and becomes the North Atlantic Current. The warm water affecting the British Isles is a branch of this current, known as the North Atlantic Drift (often incorrectly called the Gulf Stream). This branch leaves the main current in the mid-Atlantic and, passing west of Ireland, heads up towards Norway and the Arctic Ocean.

# January – In this month

**5 January 1993 –** The tanker *Braer* lost power and was wrecked on the coast of Mainland in the Shetlands. The exceptional storm of 10 January came to be called the 'Braer Storm' (see pages 28–29).

**6 January 1839 –** A major windstorm affected Ireland and Britain. It was particularly severe in Ireland, where it has come to be known as 'The Night of the Big Wind' (_Oíche na Gaoithe Móire_ in Erse). It was especially severe in northern Ireland, with gusts of over 54 m/s (100 knots or 185 km/h; 115 mph). Some 20 per cent of houses in Dublin were destroyed, and at least 250 people were killed. Much of the damage came from flooding caused by a storm surge (see page 45). The storm also severely impacted on Liverpool and northern England.

**15 January 1968 –** The Great Glasgow Storm caused considerable damage throughout the city, worse than that of 28 January 1927 (next page). Nine people were killed that night alone. The winds were extreme and a peak gust of no less than 215 kph was recorded at Great Dun Fell in Westmorland, far to the south in northern England.

**25 January 1990 –** A severe depression struck Britain. The centre tracked across southern Scotland, crossing the home of Robert Burns in Ayrshire on the anniversary of his birth, causing the event to become known as 'The Burns' Day Storm'. The most extreme winds were to the south of the depression's centre, with great damage across much of Wales and northern England. Because the worst winds struck during the day, the death toll was greater than in the 1987 October storm that affected southern England, more than twice as many (47) people being killed in Britain.

**28 January 1927** – The winds of a severe depression hit the city of Glasgow. A peak gust of 164 kph was recorded at Paisley, to the south of the city. There was widespread destruction and 11 people died, with many more injured.

**30 January 1607** – On 30 January 1607, extreme flooding hit both sides of the Bristol Channel. Cardiff was badly affected, and flood waters extended as far as Chepstow on the river Wye 3.2 km (2 miles) above where it joins the Severn. On the Devon and Somerset side, flooding was extremely extensive, with the Somerset Levels overcome. The water extended as far as Glastonbury Tor, 23 km from the Severn Estuary. It is estimated that more than 2000 individuals were drowned, a number of villages completely destroyed, and there was a widespread loss of livestock.

**31 January 1953** – A depression travelling down the North Sea caused the disastrous East Coast floods that occurred during January 31 to February 1 (some 307 lives were lost in England, 19 in Scotland, and as many as 1826 in the Netherlands). The same depression caused the loss of the *Princess Victoria* ferry in the North Channel between Scotland and Northern Ireland, with 137 lives lost in the worst British shipping disaster since the Second World War.

# The Braer Storm

On 8 January 1993 a weak wave on a front separating cold air from warmer air to its south started to develop into a depression. The centre began to move north-east but then combined with another low-pressure area, originally to its south-east. The drop in pressure was enhanced by a strong gradient in the sea-surface temperature that it encountered along its path and also by a very strong jet stream, which had winds of some 440 kph (roughly 270 mph). The system became a 'bomb', deepening by more than 1 millibar per hour for over 24 hours. By January 10 it had become the deepest extratropical cyclone ever recorded in the North Atlantic, with a central pressure estimated at 916 millibars (916 hPa). The winds associated with this depression were gale-force across the whole of the North Atlantic, and on its southern side they reached hurricane force. At its deepest, the centre was located north-west of Scotland. The ocean weather ship *Cumulus*, which was then stationed at latitude 57° 05' N, and longitude 016° 18' W (south-east of the depression's centre), recorded a maximum gust of 105 knots (194 kph, about 121 mph), and a similar gust was recorded at the remote island of North Rona, which lies 44 kilometres north-north-east of the Butt of Lewis, the northernmost point of Lewis in the Outer Hebrides. The centre continued to move north-north-east and the central pressure is estimated to have dropped to 914 hPa. Weather buoys in the area were not designed to record such low pressures, and their records only showed pressures down to 925 hPa. Gusts of 190 kph (120 mph) were measured in various locations in the north-west of the Scottish mainland. Widespread blizzards affected Scotland, when the full force of the depression reached the country.

On this occasion, the Shipping Forecast that was broadcast by the BBC gave the most extreme forecast ever issued, which was completely without precedent:

*Rockall, Malin, Hebrides, Bailey. Southwest hurricane force 12 or more.*

(Never before – or since – has the forecast used the words '*hurricane force or more*'.)

Nearly a week before, on January 5, the *MV Braer* tanker had reported loss of engine power. (Subsequent information reveals

*The oil tanker* Braer, *shortly after she ran aground at Garth's Ness near the southern tip of Mainland in Shetland. The vessel later broke up completely when hit by the ferocious storm to which the ship gave its name.*

that the vessel was technically unseaworthy, and should never have been permitted to leave Norway, bound for Canada.)

When power was lost, she was south of Sumburgh Head, the southernmost point of Mainland, the principal island in the Shetlands. Some of the members of the crew were then taken off by helicopter. Driven by the winds of a smaller, earlier depression, the vessel, despite efforts to bring her under control with a tow, grounded at Garth's Ness, near the southern tip of Mainland and immediately started to leak her cargo of crude oil (it now appears that, apart from the bad weather, poor seamanship also contributed to the disaster). When, a few days later, the vessel was hit by the force of the wind and waves of the exceptional depression that roared in from the Atlantic, the tanks were fractured, spilling the whole cargo of crude oil into the sea. Although it was a fairly light crude oil, it still caused an environmental disaster, with thousands of sea birds being killed. It is primarily for this reason that the storm has come to be known as 'the Braer Storm'. However, there was one redeeming feature: the extreme waves and winds rapidly dispersed the oil, preventing an even worse disaster.

The pressure and winds of the extreme depression began to weaken as the centre travelled north-east into the northern Atlantic and by the evening of January 10, the central pressure had risen to 920 hPa. The pressure rose to 952 hPa by the evening of January 12, and with the central pressure continuing to rise, the depression eventually decayed to the west of Norway.

February

# Introduction

Although in many rural areas (in southern England in particular), February has earned the nickname 'February Filldyke', in reality February may be extremely dry. February is, in fact, the one calendar month that is most likely to experience no rainfall whatsoever. As such, it sometimes shows spring-like conditions, which warrants its inclusion in the 'late winter, early spring' season (page 17). It has been found that a warm February in Scotland is a sign that the mean annual temperature in Scotland will probably be warm. However, a warm February may not be welcome for agriculture. Warmth indicates that precipitation will be in the form of rain, rather than snow. Rain in February is generally detrimental to seed and grass, but snow protects the ground and keeps it warm.

Yet spells of very cold weather may occur, especially if, as is often the case in January, a blocking situation occurs, especially a blocking anticyclone over Scandinavia, bringing frigid air down from the Arctic or similarly cold air from the semi-permanent cold anticyclone that builds up over Siberia in winter. The latter was the case with the exceptionally cold spell in 2018, nicknamed 'The Beast from the East' by the media, which began late in the month on February 22, and continued until at least the end of the month. This particular cold wave originated in an exceptional anticyclone, named Anticyclone Hartmut, which transported frigid Siberian air west over Europe. In crossing the North Sea, the air gained a large amount of humidity, which produced extremely heavy snowfall that spread west over almost the whole of the United Kingdom and Ireland.

Somewhat similar conditions occurred in the middle of March (March 17 and 18), although this second cold spell did not last particularly long, and was not as severe. (The media sometimes called it the 'Mini Beast from the East'.)

The weather in February often shows a division between the north and south of the country, with cold temperatures and relatively dry conditions prevailing in the north of England and over Scotland, but much milder, wetter weather in the south, sometimes with exceptional rainfalls – hence the 'February

Filldyke' nickname. The 'Beast from the East' just mentioned not only brought cold conditions to Britain, but the frigid air it brought from the east then had a great effect on the humid air introduced from the west by Storm Emma. (This storm was actually named by the meteorological services of France, Spain and Portugal, rather than the name being taken from the list prepared by the Met Office and Met Eireann.)

Storm Emma originated in the Azores and was a typical deep depression, transporting a lot of warm moist air and accompanied by high winds. When the storm encountered the frigid air introduced over the British Isles by Anticyclone Hartmut, the warm, moist air was forced up over the cold air at the surface and Emma's moisture was deposited as snowfall. The snow depth reached as much as 57 cm in places, although a depth of 50 cm was widespread across the country. South-west England and south Wales were worst affected. Storm Emma also brought a renewed incursion of Arctic air and resulting low temperatures over much of the United Kingdom.

The effects of this collision between Hartmut and Emma were not confined to the British Isles. Snow fell along the French Riviera and in Italy. Even Barcelona in Spain saw snowfall – an unheard-of event for the region. The collision of the two systems also produced some exceptional winds, most notably a gust of 228 kph (142 mph) at Mont Aigoual in southern France on 1 March 2018.

---

### Anemometer

Any device that measures wind speeds, generally in a horizontal direction. There are various types. The form most commonly seen is probably the type that has three rotating cups. Other versions use propellors, differences in pressure or the transmission of sound or heat. Certain devices (particularly sonic anemometers) are able to measure the vertical motion of the air, as well as motion in a horizontal direction.

---

# Weather Extremes

| Country | Temp. | Location | Date |
|---|---|---|---|
| *Maximum temperature* | | | |
| England | 19.7°C | Greenwich Observatory | 13 Feb. 1998 |
| Northern Ireland | 17.8°C | Bryansford (Co. Down) | 13 Feb. 1998 |
| Scotland | 17.9°C | Aberdeen (Aberdeenshire) | 22 Feb. 1897 |
| Wales | 18.7°C | Colwyn Bay (Conwy) | 23 Feb. 2012 |
| *Minimum temperature* | | | |
| England | -20.6°C | Woburn (Bedfordshire) | 25 Feb. 1947 |
| Northern Ireland | -15.0°C | Armagh (Co. Armagh) | 7 Feb. 1895 |
| Scotland | -27.2°C | Braemar (Aberdeenshire) | 11 Feb. 1895 |
| Wales | -20.0°C | Welshpool (Powys) | 2 Feb. 1954 |

| Country | Pressure | Location | Date |
|---|---|---|---|
| *Maximum pressure* | | | |
| Scotland | 1052.9 hPa | Aberdeen (Aberdeenshire) | 1 Feb. 1902 |
| *Minimum pressure* | | | |
| Eire | 942.3 hPa | Midleton (Co. Cork) | 4 Feb. 1951 |

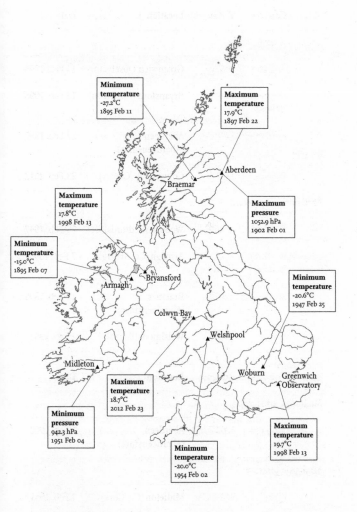

**Minimum temperature**
-27.2°C
1895 Feb 11

**Maximum temperature**
17.9°C
1897 Feb 22

Aberdeen

Braemar

**Maximum temperature**
17.8°C
1998 Feb 13

**Maximum pressure**
1052.9 hPa
1902 Feb 01

**Minimum temperature**
-15.0°C
1895 Feb 07

Armagh   Bryansford

**Minimum temperature**
-20.6°C
1947 Feb 25

Colwyn Bay

Welshpool

Midleton

Woburn   Greenwich Observatory

**Maximum temperature**
18.7°C
2012 Feb 23

**Minimum pressure**
942.3 hPa
1951 Feb 04

**Minimum temperature**
-20.0°C
1954 Feb 02

**Maximum temperature**
19.7°C
1998 Feb 13

# The Weather in February 2022

| Observation | Location | Date |
|---|---|---|
| **Max. temperature**<br>17.2°C | Pershore College<br>(Hereford & Worcester) | 16 February |
| **Min. temperature**<br>-8.1°C | Braemar (Aberdeenshire) | 11 February |
| **Rainfall**<br>86.6 mm | Seathwaite (Cumbria) | 20 February |
| **Wind gust**<br>54.5 m/s<br>(106 knots or 122 mph) | Needles Old Battery<br>(Isle of Wight) | 18 February (new<br>low-level record) |
| **Snow depth**<br>12 cm | Aviemore<br>(Inverness-shire) | 10 February |

*Trade winds*
There are two trade-wind zones, north and south of the equator, with the north-east trades and the south-east trades, respectively, where the air converges on the low-pressure region at the equator. The direction and strength of these winds do remain relatively constant throughout the year, and were thus a reliable source of motive power for sailing ships. The term 'trade' actually derives from the constancy of the winds, and not from their importance commercially.

February 2022 was notable in the period in the middle of the month for the extreme winds that affected the country. There were no less than three named storms: Dudley, Eunice and Franklin.

Although the month started quietly, with mild weather almost everywhere, there was considerable snow in Scotland (Braemar recorded a minimum temperature of -8.1°C on February 11). Overall there was more rain than average in the north of England (but below average in English counties along the south coast). Both Topcliffe (in North Yorkshire) and Redesdale (in Northumberland) recorded overnight temperatures of -7.8°C on February 10.

Storm Dudley arrived on February 16 and caused major travel disruption and power outages in Scotland and northern England. There were also power outages in eastern England and roads blocked in the south-east.

The second storm, Storm Eunice (see pages 44 and 45), swept in on February 18. It caused major damage across a wide swathe of the country. Snowfall in Scotland caused travel problems, and there was severe flooding in Northern Ireland. There was widespread disruption to travel in Wales, and even in southern England. A 'Do not travel' warning was issued for the London area and the London mayor urged people to stay at home. Some ferry crossings to and from France at Dover were cancelled. The wind gust of 106 knots (122 mph) recorded at Needles Old Battery (Isle of Wight) during Storm Eunice on February 18 set a new low-level record.

The third storm, Storm Franklin, affected Northern Ireland with road closures, flooding and some wind damage. North-west England was also badly affected, but the storm caused problems right across England. There was flooding in Yorkshire. Train travel was particularly badly affected in northern England and the Midlands, where many roads were flooded. In southern England, the railways issued a 'Do not travel' warning.

Well over half a million customers had power restored after the two storms, Eunice and Franklin.

# Sunrise and Sunset 2023

| Location | Date | Rise | Azimuth ° | Set | Azimuth ° |
|---|---|---|---|---|---|
| **Belfast** | | | | | |
| | Feb 01 (Wed) | 08:13 | 119 | 17:02 | 241 |
| | Feb 11 (Sat) | 07:54 | 114 | 17:22 | 247 |
| | Feb 21 (Tue) | 07:33 | 107 | 17:43 | 253 |
| | Feb 28 (Tue) | 07:16 | 103 | 17:57 | 258 |
| **Cardiff** | | | | | |
| | Feb 01 (Wed) | 07:51 | 117 | 17:02 | 243 |
| | Feb 11 (Sat) | 07:35 | 112 | 17:20 | 248 |
| | Feb 21 (Tue) | 07:15 | 106 | 17:38 | 254 |
| | Feb 28 (Tue) | 07:01 | 102 | 17:51 | 258 |
| **Edinburgh** | | | | | |
| | Feb 01 (Wed) | 08:08 | 120 | 16:45 | 240 |
| | Feb 11 (Sat) | 07:47 | 114 | 17:07 | 246 |
| | Feb 21 (Tue) | 07:24 | 108 | 17:29 | 252 |
| | Feb 28 (Tue) | 07:07 | 103 | 17:44 | 257 |
| **London** | | | | | |
| | Feb 01 (Wed) | 07:40 | 117 | 16:50 | 243 |
| | Feb 11 (Sat) | 07:23 | 112 | 17:08 | 248 |
| | Feb 21 (Tue) | 07:04 | 106 | 17:26 | 254 |
| | Feb 28 (Tue) | 06:49 | 102 | 17:39 | 258 |

*Note that all times are in Universal Time (UT), otherwise known as Greenwich Mean Time (GMT). These times do not take Summer Time (BST) into account.*

# Moonrise and Moonset 2023

| Location | Date | Rise | Azimuth ° | Set | Azimuth ° |
|----------|------|------|-----------|-----|-----------|
| **Belfast** | | | | | |
| | Feb 01 (Wed) | 12:17 | 038 | 06:07 | 321 |
| | Feb 11 (Sat) | – | – | 09:43 | 254 |
| | Feb 21 (Tue) | 08:29 | 105 | 19:29 | 260 |
| | Feb 28 (Tue) | 10:14 | 039 | 03:58 | 320 |
| **Cardiff** | | | | | |
| | Feb 01 (Wed) | 12:31 | 043 | 05:31 | 316 |
| | Feb 11 (Sat) | 23:52 | 110 | 09:37 | 256 |
| | Feb 21 (Tue) | 08:12 | 104 | 19:21 | 261 |
| | Feb 28 (Tue) | 10:27 | 044 | 03:23 | 315 |
| **Edinburgh** | | | | | |
| | Feb 01 (Wed) | 11:54 | 036 | 06:08 | 323 |
| | Feb 11 (Sat) | – | – | 09:29 | 254 |
| | Feb 21 (Tue) | 08:20 | 106 | 19:16 | 260 |
| | Feb 28 (Tue) | 09:51 | 037 | 03:59 | 322 |
| **London** | | | | | |
| | Feb 01 (Wed) | 12:18 | 043 | 05:20 | 316 |
| | Feb 11 (Sat) | 23:40 | 110 | 09:25 | 256 |
| | Feb 21 (Tue) | 08:01 | 104 | 19:09 | 261 |
| | Feb 28 (Tue) | 10:14 | 044 | 03:11 | 315 |

*Note that all times are in Universal Time (UT), otherwise known as Greenwich Mean Time (GMT). These times do not take Summer Time (BST) into account.*

# Twilight Diagrams 2023

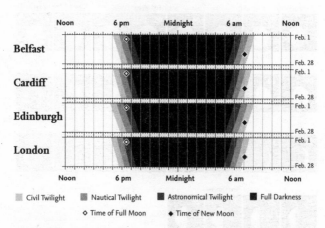

| | |
|---|---|
| ▨ Civil Twilight | ▨ Nautical Twilight | ■ Astronomical Twilight | ■ Full Darkness |
| ◇ Time of Full Moon | ◆ Time of New Moon | | |

The exact times of the Moon's major phases are shown on the diagrams opposite.

### Front

A zone separating two air masses with different characteristics (typically, with different temperatures and/or humidities). Depressions normally show two fronts: a warm front (where warm air is advancing) and a cold front (where cold air is advancing). The latter normally move faster than warm fronts. When a cold front catches up with a warm front, the warm air is lifted away from the surface, giving a pool of warm air at altitude. The combined front is known as an occluded front. Depending on the exact conditions, occluded fronts may give long periods of overcast skies and persistent rain.

# The Moon's Phases and Ages 2023

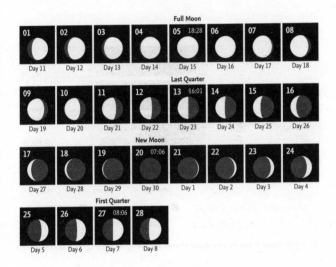

**Full Moon**

| 01 | 02 | 03 | 04 | 05 18:28 | 06 | 07 | 08 |
|---|---|---|---|---|---|---|---|
| Day 11 | Day 12 | Day 13 | Day 14 | Day 15 | Day 16 | Day 17 | Day 18 |

**Last Quarter**

| 09 | 10 | 11 | 12 | 13 16:01 | 14 | 15 | 16 |
|---|---|---|---|---|---|---|---|
| Day 19 | Day 20 | Day 21 | Day 22 | Day 23 | Day 24 | Day 25 | Day 26 |

**New Moon**

| 17 | 18 | 19 | 20 07:06 | 21 | 22 | 23 | 24 |
|---|---|---|---|---|---|---|---|
| Day 27 | Day 28 | Day 29 | Day 30 | Day 1 | Day 2 | Day 3 | Day 4 |

**First Quarter**

| 25 | 26 | 27 08:06 | 28 |
|---|---|---|---|
| Day 5 | Day 6 | Day 7 | Day 8 |

*Cyclone*
Technically, a name for any circulation of air around a low-pressure centre. (Depressions are also known as 'extratropical cyclones'.) The term is also used specifically for a tropical, revolving storm in the Indian Ocean, known as a 'hurricane' over the North Atlantic Ocean or eastern Pacific Ocean. The term 'typhoon' is used for systems in the western Pacific that affect northern Australia and Asia. The term 'tropical cyclones' applies to all such revolving systems.

# February – In this month

**31 January to 1st February 1953** – North Sea flood of 1953 continued to affect eastern coastal counties in the beginning of the month.

**2 February (Candlemass)** – A German superstition arose around a hedgehog casting a shadow on Candlemass day (February 2). This was the origin of the North American tradition of Groundhog Day, when, if a groundhog emerges from its burrow and sees its shadow, winter will persist for another six weeks. If there is no shadow, spring will arrive early.

**2 February 1602** – A particularly hard frost affected the Faroe Islands on Candlemass that year. It is still known as 'the Hard Candlemass' in the islands.

**4 February 1784** – The eruption of the Laki volcanic fissure, which caused 'dry fog' and crop failures across Europe, finally ended. The eruption had begun in June 1783.

**4 February 1968** – Off Iceland, the trawler Ross Cleveland was overwhelmed by ice accumulation and sank, with just one member of crew surviving.

**18 February 2022** – The very destructive Storm Eunice affected the British Isles (see pages 44 and 45).

**25 February 1947** – There was extreme snowfall from January to March 1947. February 25 saw the greatest single snowfall on record that lasted for close to 50 consecutive hours.

**26 February 2019** – The highest temperature for any winter month was the 21.2°C recorded at Kew Gardens (London).

# Storm Eunice

Of the three named storms that affected the British Isles in February 2022, it was probably Storm Eunice that caused the most widespread disruption and damage. In an almost unprecedented step, the Met Office issued a red warning (the most extreme) of winds for south Wales and south-west England on February 17, and followed this on February 18 with a red warning for London and south-east England. The storm arrived on February 18. Heavy snowfall in Scotland disrupted travel in that region, particularly on the roads. Ferry services to Larne in Northern Ireland from Cairnryan in Ayrshire were cancelled. There were school closures in Aberdeenshire and Angus.

There was severe flooding in Northern Ireland, and some travel problems, particularly with fallen trees and snow blocking roads. There were power outages for a few hundred customers. There was slightly less disruption in the area than with the previous named storms that arrived in February, Storms Dudley and Franklin.

In Wales, there was both road and rail disruption, mainly caused by fallen trees, but only minor damage to buildings and few injuries, mainly thanks to action by the Welsh Government, which ordered the pre-emptive closure of many schools, colleges and public buildings. There was also pre-emptive cancellation of many bus and train services, and both ferry and flights were delayed. Both Severn bridges (and other bridges) were closed. Tens of thousands of customers lost electrical power.

In England there were major problems in the north-west with power outages to tens of thousands of customers. Both road and rail links were blocked by fallen trees, with many train cancellations. Trains were particularly badly affected in the Preston and Manchester areas, with both delays and cancellations. Trains were also affected in eastern England, with many cancellations. In eastern England about 1000 customers suffered a loss of power and there were many school closures.

In London, the QEII bridge across the Thames was closed. There were delays and cancellations to overground rail movements and the rail operators issued a 'Do not travel' warning. The Mayor of London urged everyone to stay at home. Many buildings were damaged by the high winds.

In the south-east the situation was also severe, with many train cancellations. Many roads were blocked by fallen trees and there was some damage to buildings along the south coast. Ferry crossings to France from Dover were cancelled. In West Sussex, 200-year-old cedar trees that had survived the Great Storm of 15–16 October 1987 were toppled by the high winds of Storm Eunice. The gust of 106 knots (122 miles/hr or 196 km/hr) recorded at the Needles Old Battery on the Isle of Wight is a new, provisional, record for a low-level gust. The same station, the Needles Old Battery, had recorded a gust of 86 knots (97 miles/hr or 159 km/hr) the previous year, on 13 March 2021, and one of 81 knots (93 miles/hr or 150 km/hr) on 3 May 2021.

A similar situation prevailed in south-western counties, with disruption to road, rail and air travel. There were cancellations to flights from both Bristol and Exeter airports.

---

### Storm surge

A raised level of seawater that is driven ashore and may cause flooding many kilometres inland. It is a major danger of tropical cyclones that make landfall. The level of the sea is raised primarily by the lower atmospheric pressure at the centre of depressions or tropical cyclones, and may be increased by a high tide (especially a spring tide). The water is often driven ashore by onshore winds. Storm surges have frequently caused many thousands of deaths, especially in the most vulnerable, low-lying coastal areas of India and Bangladesh.

---

March

# March

# Introduction

March has traditionally been viewed as a transitional month between winter and spring. It has come to be associated with the saying 'In like a lion, out like a lamb'. Regrettably for this saying, although the weather may well be changeable, there is no distinct pattern that applies every year. The month may well begin with a settled period, with little in the way of winds and dramatic events, and then deteriorate. Interestingly, there is often a sharp change in the weather after the first week of March. The frequency of westerly weather – that is, the progression of depressions across the country from the west – drops suddenly, and very few are recorded. The number of such depressions and sequences of weather is less than at any other time of the year, except for late April and early May. In general, the increasing strength of heating from the Sun begins to take effect from the beginning of the month. Although there may still be spells of very cold weather, with a lot of cloud, the increased warmth is readily noticed.

Another tradition, strongly held by some, particularly by mariners, is that March experiences 'equinoctial gales'. (The equinox is on March 20 in 2023.) This idea is, however, not supported by the evidence. Gales are actually most common around the winter solstice (December 22 in 2023), and least frequent around the summer solstice (June 21). There may be a perception that gales are most frequent around the autumnal equinox (September 23 in 2023), but this is probably a result of the deterioration of the weather from the fine, long days of summer to the shorter, less settled days of autumn. When the Sun is well north of the equator, its warmth generally causes the Azores High to strengthen and extend its influence as ridges of high pressure. These may reach so far as to cover the British Isles and most of western Europe. Sometimes the high-pressure region extends well to the west, giving rise to what is known as the Bermuda High. Depressions, and their accompanying cloud and winds, tend to follow the Polar Front jet stream, which is diverted towards the north. The low-pressure systems, with their clouds, rain and winds, pass north of the British Isles, which thus remains under the influence of the settled, anticyclonic weather within the extended Azores High. With the change of season from summer to autumn, the Sun starts to move south. It moves fastest towards the equator and south of it in September, usually causing the high-pressure area over the Azores to decline. When this happens, depressions are no longer displaced northwards and they, and their accompanying winds, thus cross the British Isles more directly. It would seem that mariners' observations of such windier weather in September have been taken in a more general sense to suggest that gales occur at both equinoxes (in March and September). There is no evidence whatsoever for increased windiness at either equinox, so this is yet another instance of incorrect weather lore.

---

### Azores High

A more-or-less permanent high-pressure system in the North Atlantic, generally centred approximately over the islands of the Azores, or closer to Iberia (Portugal and Spain). It arises from air that has risen at the equator that descends at the sub-tropical high-pressure zones.

---

# Weather Extremes

| Country | Temp. | Location | Date |
|---------|-------|----------|------|
| *Maximum temperature* | | | |
| England | 25.6°C | Mepal (Cambridgeshire) | 29 Mar. 1968 |
| Northern Ireland | 21.7°C | Armagh (Co. Armagh) | 28 Mar. 1965 29 Mar. 1965 |
| Scotland | 23.6°C | Aboyne (Aberdeenshire) | 27 Mar. 2012 |
| Wales | 23.0°C | Prestatyn (Denbighshire) Ceinws (Powys) | 29 Mar. 1965 |
| *Minimum temperature* | | | |
| England | -21.1°C | Houghall (Co. Durham) | 4 Mar. 1947 |
| Northern Ireland | -14.8°C | Katesbridge (Co. Down) | 2 Mar. 2001 |
| Scotland | -22.8°C | Logie Coldstone (Aberdeenshire) | 14 Mar. 1958 |
| Wales | -21.7°C | Corwen (Denbighshire) | 3 Mar. 1965 |

| Country | Pressure | Location | Date |
|---------|----------|----------|------|
| *Maximum pressure* | | | |
| England | 1047.9 hPa | St Mary's Airport (Isles of Scilly) | 9 Mar. 1953 |
| *Minimum pressure* | | | |
| Scotland | 946.2 hPa | Wick (Caithness) | 9 Mar. 1896 |

Minimum
pressure
946.2 hPa
1896 Mar 09

Minimum
temperature
-22.8°C
1958 Mar 14

Maximum
temperature
23.6°C
2012 Mar 27

Wick

Logie Coldstone
Aboyne

Minimum
temperature
-14.8°C
2001 Mar 02

Minimum
temperature
-21.1°C
1947 Mar 04

Maximum
temperature
21.7°C
1965 Mar 28
1965 Mar 29

Houghall

Armagh   Katesbridge

Maximum
temperature
23.0°C
1965 Mar 29

Prestatyn
Corwen

Ceinws

Mepal

Maximum
temperature
23.0°C
1965 Mar 29

Maximum
temperature
25.6°C
1968 Mar 29

St Mary's Airport

Maximum
pressure
1047.9 hPa
1953 Mar 09

Minimum
temperature
-21.7°C
1965 Mar 03

# The Weather in March 2022

| Observation | Location | Date |
|---|---|---|
| *Max. temperature* 20.8°C | St James's Park (London) Treknow (Cornwall) | 25 March |
| *Min. temperature* -9.1°C | Aboyne (Aberdeenshire) | 2 March |
| *Rainfall* 51.6 mm | White Barrow (Devon) | 2 March |
| *Wind gust* 32.4 m/s (63 knots or 72 mph) | St Mary's (Scilly) | 12 March |
| *Sunshine* 11.7 hrs | Shoeburyness (Essex) | 26 March |
| *Snow depth* 10 cm | Copley (Durham) | 31 March |

The month of March 2022 began with cold weather over the whole country, with frosty conditions in Scotland, where the month's minimum temperature of -9.1°C was recorded at Aboyne in Aberdeenshire on March 2. In south-eastern England it was mild, with rain spreading in from the west and crossing the country, for the first few days of the month. Some of this rain was heavy. White Barrow in Devon recorded the month's maximum rainfall of 51.6 mm on March 2. This rain meant that the month's rainfall was above average in the south-west and in southern counties, although low elsewhere, particularly in north-western Scotland, which was notably dry. Some of the rain was accompanied by high winds, with St Mary's in the Scilly Isles reporting 32.4 m/s (63 knots or 72 mph) on March 12.

Following the unsettled conditions at the beginning of the month, from the middle of the month, anticyclonic conditions over the whole country brought clear skies during the daytime, but chilly nights, which were especially cold over northern Scotland.

It was a sunny month, provisionally the sunniest over the whole country since 1929. Both Scotland and Northern Ireland recorded the sunniest March since 1919. During March 19 to 22 there were localised outbreaks of wildfires in Northern Ireland, northern England and Wales, almost certainly exacerbated by the strong, drying winds.

An intense high-pressure area then arose, centred near Denmark. Temperatures in north-western Scotland became very high, thanks to a föhn effect (see page 136) and 20.2°C was recorded at Kinlochewe (Highland). The next few days saw high temperatures everywhere and Wales experienced 20.7°C at Porthmadog in Gwynedd on March 22, with the highest of 20.8°C at St James's Park in Central London on March 23. Treknow in Cornwall reached the same temperature on March 25.

At the very end of the month the high pressure had migrated to become centred over the UK, but this allowed an incursion of Arctic air. This gradually brought colder, more wintry conditions to the whole country over the next few days with particularly heavy snowfall in the eastern counties of northern England. There were considerable falls of graupel (snow pellets) and snow. Copley, in Durham, reported a snow depth of 10 cm on March 31.

# Sunrise and Sunset 2023

| Location | Date | Rise | Azimuth ° | Set | Azimuth ° |
|---|---|---|---|---|---|
| **Belfast** | | | | | |
| | Mar 01 (Wed) | 07:14 | 102 | 17:59 | 258 |
| | Mar 11 (Sat) | 06:49 | 95 | 18:19 | 265 |
| | Mar 21 (Tue) | 06:24 | 89 | 18:39 | 272 |
| | Mar 31 (Fri) | 05:59 | 82 | 18:58 | 279 |
| **Cardiff** | | | | | |
| | Mar 01 (Wed) | 06:59 | 101 | 17:53 | 259 |
| | Mar 11 (Sat) | 06:37 | 95 | 18:10 | 265 |
| | Mar 21 (Tue) | 06:14 | 89 | 18:27 | 272 |
| | Mar 31 (Fri) | 05:51 | 82 | 18:44 | 278 |
| **Edinburgh** | | | | | |
| | Mar 01 (Wed) | 07:05 | 102 | 17:46 | 258 |
| | Mar 11 (Sat) | 06:39 | 94 | 18:07 | 265 |
| | Mar 21 (Tue) | 06:13 | 89 | 18:28 | 272 |
| | Mar 31 (Fri) | 05:47 | 82 | 18:48 | 279 |
| **London** | | | | | |
| | Mar 01 (Wed) | 06:47 | 101 | 17:41 | 259 |
| | Mar 11 (Sat) | 06:25 | 95 | 17:58 | 265 |
| | Mar 21 (Tue) | 06:02 | 89 | 18:15 | 272 |
| | Mar 31 (Fri) | 05:39 | 82 | 18:32 | 278 |

*Note that all times are in Universal Time (UT), otherwise known as Greenwich Mean Time (GMT). These times do not take Summer Time (BST) into account.*

# Moonrise and Moonset 2023

| Location | Date | Rise | Azimuth ° | Set | Azimuth ° |
|----------|------|------|-----------|-----|-----------|
| **Belfast** | | | | | |
| | Mar 01 (Wed) | 10:57 | 37 | 05:03 | 323 |
| | Mar 11 (Sat) | 23:23 | 119 | 08:01 | 247 |
| | Mar 21 (Tue) | 06:45 | 99 | 18:25 | 266 |
| | Mar 31 (Fri) | 12:00 | 46 | 04:57 | 316 |
| **Cardiff** | | | | | |
| | Mar 01 (Wed) | 11:12 | 42 | 04:25 | 318 |
| | Mar 11 (Sat) | 23:00 | 117 | 07:58 | 249 |
| | Mar 21 (Tue) | 06:31 | 99 | 18:15 | 266 |
| | Mar 31 (Fri) | 12:08 | 50 | 04:25 | 312 |
| **Edinburgh** | | | | | |
| | Mar 01 (Wed) | 10:32 | 34 | 05:05 | 326 |
| | Mar 11 (Sat) | 23:17 | 120 | 07:46 | 346 |
| | Mar 21 (Tue) | 06:36 | 100 | 18:13 | 266 |
| | Mar 31 (Fri) | 11:39 | 44 | 04:55 | 318 |
| **London** | | | | | |
| | Mar 01 (Wed) | 10:58 | 42 | 04:14 | 318 |
| | Mar 11 (Sat) | 22:48 | 117 | 07:46 | 249 |
| | Mar 21 (Tue) | 06:19 | 99 | 18:02 | 166 |
| | Mar 31 (Fri) | 11:55 | 50 | 04:14 | 312 |

*Note that all times are in Universal Time (UT), otherwise known as Greenwich Mean Time (GMT). These times do not take Summer Time (BST) into account.*

# Twilight Diagrams 2023

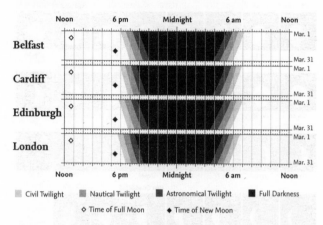

The exact times of the Moon's major phases are shown on the diagrams opposite.

---

### Air mass

A large volume of air that has uniform properties (particularly temperature and humidity) throughout. Air masses arise when air stagnates over a particular area for a long time. These areas are known as 'source regions' and are generally the semi-permanent high-pressure zones, which are the sub-tropical and polar anticyclones. The primary classification is based on temperature, giving Arctic (A), polar (P) and tropical (T) air.

---

# The Moon's Phases and Ages 2023

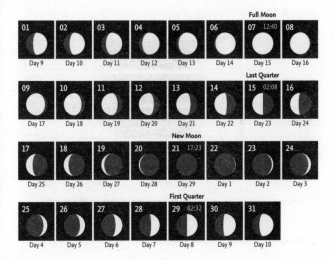

---

*Lapse rate*
The change in a property with increasing altitude. In meteorology, this is usually the change in temperature. In the troposphere (the lowest layer of the atmosphere), this is a decrease in temperature with an increase in height. This is defined as a positive lapse rate. In the stratosphere (the next higher layer) there is an overall increase in temperature with height, giving a negative lapse rate.

---

# March – In this month

**2 March 1963** – A family on Dartmoor is finally rescued after 65 days marooned in a remote farmhouse among 8-metre drifts of snow.

**4 March 1947** – A blizzard sweeps in from the west and buries England and Wales in snow. Snow has fallen somewhere every day for six weeks. It is the snowiest winter since records began.

**13 March 1947** – The 'biggest road transport hold-up ever known' occurs today with vehicles at a standstill as a result of a very heavy snowfall on Shap Fell in Westmoreland (now in Cumbria). With a height of 486 metres it was the highest main road in Britain, and the principal route to Scotland, until a lower motorway was opened in 1970.

**16 March 1947** – The 'Great Suck' begins. Thawing of the enormous snowfall leads to breaches in the dykes in the Fens and widespread flooding. Local authorities, the fire brigade and the army begin to use pumps to reduce the flooding. Some land is not freed of lying water until June.

**21 March 1748** – A violent Atlantic storm, encountered during a passage home, 20 years later, inspires John Newton (together with the poet William Cowper) to write 'Amazing Grace', still the most popular hymn in Britain.

**23 March 1950** – the World Meteorological Organization was founded (see pages 60 and 61).

**23 March 1961** – World Meteorology Day established in 1961 to commemorate the formation of the World Meteorological Organisation.

**24 March 1878** – The sail-training ship, *HMS Euridice*, is hit by a sudden, blinding snowstorm and squall off Sandown Bay in the Isle of Wight. The snow is so thick that the crew could not see to strike the sails. The ship keels over and sinks. All but two of the 370 on board are drowned. Within minutes the squall and snow shower pass.

**27–28 March 1916** – A widespread severe northerly gale and associated blizzard affected much of East Anglia, the east and south Midlands, parts of south-east England and the west of England. Large numbers of trees were lost. (Kew Observatory reported Force 11 for a short time in the early evening of March 28.) In some places the snowfall lasted more than 24 hours. The combination of rain and snow surpassed 50–60 mm over a wide swathe from Cornwall to Norfolk.

**30 March 1930** – The steam trawler *Ben Doran* was wrecked in storm-force winds on the Ve Skerries, Shetland. The whole crew, estimated to be nine in number, were killed. The wreck has been described as 'the most tragic wreck in all Shetland's history'. The Stromness (Orkney) lifeboat was launched, but arrived far too late to affect a rescue. The tragedy led to the establishment of the first lifeboat station on Shetland.

# World Meteorological Organization

**The World Meteorological Organization**

The World Meteorological Organization (WMO) is a specialised agency of the United Nations devoted to meteorology and related sciences. It was established on 23 March 1950 as a formalised successor to the International Meteorological Organization (IMO), which had been set up in September 1873, in recognition of the fact that weather systems crossed national boundaries and that co-operation in the study of atmospheric characteristics and the exchange of data was essential to the practice of successful weather forecasting. The IMO itself originated in a conference held in Cambridge as early as 1845. Preparations for a formal agreement were not made until September 1873.

March 23 (the day of the WMO's formation), is celebrated yearly as World Meteorology Day. The first such day was held on 23 March 1961. The WMO issues a slogan for every Meteorological Day and some recent slogans have been:

*Early Warning and Early Action (2022)*
*The Ocean, Our Climate and Weather (2021)*
*Climate and Water (2020)*
*The Sun, the Earth and the Weather (2019)*
*Weather-ready, Climate-smart (2018)*
*Understanding Clouds (2017)*
*Hotter, Drier, Wetter. Face the Future (2016)*
*Climate Knowledge for Climate Action (2015)*
*Weather and Climate: Engaging Youth (2014)*
*Watching the Weather to Protect Life and Property: Celebrating 50 Years of World Weather Watch (2013)*

The World Weather Watch mentioned in the last of these slogans is the coordinated, worldwide system that ensures that all member states obtain the data and services required for weather forecasting and research. It consists of three main elements: the Global Observing System, the Global Data-Processing System, and the Global Telecommunications System. Although described separately, these three elements function as a unified whole, and are devoted to obtaining standardised meteorological data, wherever obtained; maintaining this data at specific centres, and making it available to anyone that requires it; and ensuring that the communications systems are in place for any organisation to obtain the data that it requires.

The standardization of data is extremely important, and the WMO issues recommendations as to the way in which observations are made and reported to ensure data is universally intelligible.

The WMO consists of 193 nations and territories, divided into various regions. Two nations (France and the United Kingdom) are members of four regions each, and ten other states are members of two regions.

**Region I**
Africa

**Region II**
Middle East and Southern Asia

**Region III**
South America

**Region IV**
Canada and Carribean

**Region V**
Oceania

**Region VI**
Central Europe

**Multiple Regions**
UK, USA, Russia, Western Europe, Venezuela

*For forecasting purposes, the WMO groups nations by regions. A few countries handle several regions.*

April

# Introduction

April is traditionally associated with a changeable month and the general view is that it is accompanied by 'April showers'. Although it is certainly changeable, with the transition from winter to the warmer conditions of summer, in recent years the showers have been less frequent. Certainly, there are normally periods when northerly winds (in particular) bring showers and these may be squally, and can sometimes even turn into thunderstorms and be accompanied by hail. Generally nowadays, the showers are infrequent, and it is one of the quietest times of the year. April is actually one of the driest months for most of the country, followed in sequence by March, June, May and February.

The temperature of the sea surrounding the British Isles remains low, while the land begins to warm up. Sea breezes are common and they are often accompanied by sea fog. This is particularly frequent on the east coast of Britain. Sea fogs often form over the cold North Sea and then are carried inland by the sea breezes that build up during the day. In Scotland, in particular, the 'haar', as it is known, often moves in during the afternoon. It is often particularly noticeable when it invades the Firth of Forth and hides the bottom of the Forth Bridges from view. A similar phenomenon occurs anywhere along the east coast of England, particularly in Northumbria, but is found as far south as East Anglia. The same sort of sea fog does occur on the western coasts of Britain, but is less common, because the sea surface temperature of the Atlantic water tends to be higher than that of the North Sea, so sea fog is less likely to form.

### *Spring and early summer season – April 1 to June 17*

The weather during this season is very changeable. Indeed, it is the most changeable of the year. There are occasional outbreaks of northerly winds, which tend to produce heavy squally showers, often turning into thunderstorms with lightning and even hail. Initially, high pressure and dry air over Continental Europe often extends west over the British Isles, but this high pressure and its accompanying dry, continental air tend to collapse in early summer and be replaced by moist maritime air as depressions move across the country from the west, on tracks towards the Baltic or slightly farther north, towards Scandinavia.

# Weather Extremes

| Country | Temp. | Location | Date |
|---------|-------|----------|------|
| *Maximum temperature* | | | |
| England | 29.4°C | Camden Square (London) | 16 Apr. 1949 |
| Northern Ireland | 24.5°C | Boom Hall (Co. Londonderry) | 26 Apr. 1984 |
| Scotland | 27.3°C | Inverailort (Highland) | 17 Apr. 2003 |
| Wales | 26.2°C | Gogerddan (Ceredigion) | 16 Apr. 2003 |
| *Minimum temperature* | | | |
| England | -15°C | Newton Rigg (Cumbria) | 2 Apr. 1917 |
| Northern Ireland | -8.5°C | Killylane (Co. Antrim) | 10 Apr. 1998 |
| Scotland | -13.3°C | Braemar (Aberdeenshire) | 11 Apr. 1917 |
| Wales | -11.2°C | Corwen (Denbighshire) | 11 Apr. 1978 |

| Country | Pressure | Location | Date |
|---------|----------|----------|------|
| *Maximum pressure* | | | |
| Scotland | 1044 hPa | Eskdalemuir (Dumfrieshire) | 11 Apr. 1938 |
| *Minimum pressure* | | | |
| Northern Ireland | 952.9 hPa | Malin Head (Co. Donegal) | 1 Apr. 1948 |

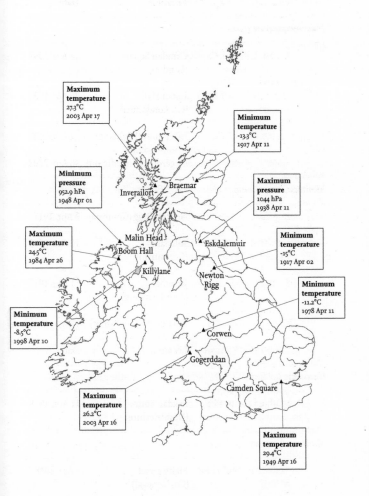

**Maximum temperature**
27.3°C
2003 Apr 17

**Minimum temperature**
-13.3°C
1917 Apr 11

**Minimum pressure**
952.9 hPa
1948 Apr 01

**Maximum pressure**
1044 hPa
1938 Apr 11

**Maximum temperature**
24.5°C
1984 Apr 26

**Minimum temperature**
-15°C
1917 Apr 02

**Minimum temperature**
-11.2°C
1978 Apr 11

**Minimum temperature**
-8.5°C
1998 Apr 10

Braemar

Inverailort

Malin Head
Boom Hall
Killylane

Eskdalemuir

Newton Rigg

Corwen

Gogerddan

Camden Square

**Maximum temperature**
26.2°C
2003 Apr 16

**Maximum temperature**
29.4°C
1949 Apr 16

67

# The Weather in April 2022

| Observation | Location | Date |
| --- | --- | --- |
| *Max. temperature* 23.4°C | St James's Park (Central London) | 15 April |
| *Min. temperature* -8.0°C | Tulloch Bridge (Inverness-shire) | 1 April |
| *Rainfall* 62.4 mm | Achfary (Sutherland) | 4 April |
| *Wind gust* 33.4 m/s (65 knots or 75 mph) | Needles Old Battery (Isle of Wight) | 7 April |
| *Snow depth* 10 cm | Redesdale Camp (Northumberland) | 1 April |

### *Nacreous clouds*

Brilliantly coloured clouds (also known as 'mother-of-pearl' clouds) that are occasionally seen at sunset or sunrise. They occur in the lowest region of the stratosphere at altitudes of 15–30 kilometres. They arise when wave motion at altitude causes water vapour to freeze onto suitable nuclei at very low temperatures (below -83°C).

The cold air that invaded the UK at the end of March gave wintry conditions early in April 2022. The minimum temperature for the month was -8.0°C at Tulloch Bridge (Inverness-shire) on April 1. The same day saw a snow depth of 10 cm at Redesdale Camp in Northumberland.

Initially, the weather was unsettled and cold, with temperatures rising in the middle of April. At this time, anticyclonic conditions became established over most of the country, with clear skies giving sunny days but widespread frosts at night. Sennybridge in Powys had its lowest April temperature for nine years with -7.5°C on April 3. There were distinct differences in temperature between the two sides of the frontal zone that marked the extent of the Arctic air. On April 5, for example, the temperature was 11°C over Scotland's Central Belt, but just 1.4°C at Aboyne in Aberdeenshire.

A depression tracked its way across the country on April 7, giving rise to some high winds on its southern side. The maximum gust for the month was 33.4 m/s (65 knots or 75 mph), recorded at the Needles Old Battery on the Isle of Wight on April 7.

Early in the month there was considerable rain in northern Scotland, with the maximum rainfall for the month of 62.4 mm occurring in the 24 hours to 9:00 am on April 4 at Achfary in Sutherland. With the exceptions of northern Scotland and parts of Northern Ireland, in other regions the month of April was relatively dry.

The anticyclonic conditions that became established during the middle of April meant that the Easter weekend was warm and dry (Easter Sunday was April 17). The month's maximum temperature of 23.4°C was recorded at St James's Park in Central London on Good Friday, April 15.

Towards the end of the month, conditions were generally very quiet, but widespread cloud gave subdued temperatures over most of the country, although clear skies at night gave sharp frosts in certain sheltered north-western areas. The month ended with Atlantic fronts pushing in from the west, giving wet conditions in Northern Ireland and Scotland.

# Sunrise and Sunset 2023

| Location | Date | Rise | Azimuth ° | Set | Azimuth ° |
|----------|------|------|-----------|-----|-----------|
| **Belfast** | | | | | |
| | Apr 01 (Sat) | 05:56 | 81 | 19:00 | 279 |
| | Apr 11 (Tue) | 05:31 | 74 | 19:19 | 286 |
| | Apr 21 (Fri) | 05:08 | 68 | 19:39 | 292 |
| | Apr 30 (Sun) | 04:48 | 63 | 19:56 | 298 |
| **Cardiff** | | | | | |
| | Apr 01 (Sat) | 05:48 | 82 | 18:46 | 279 |
| | Apr 11 (Tue) | 05:26 | 75 | 19:03 | 285 |
| | Apr 21 (Fri) | 05:05 | 70 | 19:19 | 291 |
| | Apr 30 (Sun) | 04:47 | 65 | 19:34 | 296 |
| **Edinburgh** | | | | | |
| | Apr 01 (Sat) | 05:44 | 81 | 18:51 | 280 |
| | Apr 11 (Tue) | 05:18 | 74 | 19:11 | 287 |
| | Apr 21 (Fri) | 04:53 | 67 | 19:31 | 293 |
| | Apr 30 (Sun) | 04:32 | 62 | 19:50 | 299 |
| **London** | | | | | |
| | Apr 01 (Sat) | 05:37 | 82 | 18:34 | 279 |
| | Apr 11 (Tue) | 05:14 | 75 | 18:51 | 285 |
| | Apr 21 (Fri) | 04:53 | 70 | 19:08 | 291 |
| | Apr 30 (Sun) | 04:35 | 65 | 19:23 | 296 |

*Note that all times are in Universal Time (UT), otherwise known as Greenwich Mean Time (GMT).*

# Moonrise and Moonset 2023

| Location | Date | Rise | Azimuth ° | Set | Azimuth ° |
|---|---|---|---|---|---|
| **Belfast** | | | | | |
| | Apr 01 (Sat) | 13:17 | 54 | 05:15 | 308 |
| | Apr 11 (Tue) | 01:30 | 142 | 07:32 | 217 |
| | Apr 21 (Fri) | 05:33 | 60 | 21:51 | 307 |
| | Apr 30 (Sun) | 13:31 | 63 | 03:47 | 295 |
| **Cardiff** | | | | | |
| | Apr 01 (Sat) | 13:20 | 57 | 04:48 | 305 |
| | Apr 11 (Tue) | 00:53 | 137 | 07:47 | 222 |
| | Apr 21 (Fri) | 05:33 | 62 | 21:24 | 303 |
| | Apr 30 (Sun) | 03:28 | 70 | 03:26 | 293 |
| **Edinburgh** | | | | | |
| | Apr 01 (Sat) | 12:58 | 53 | 05:12 | 310 |
| | Apr 11 (Tue) | 01:32 | 144 | 07:08 | 215 |
| | Apr 21 (Fri) | 05:16 | 59 | 21:47 | 308 |
| | Apr 30 (Sun) | 13:16 | 68 | 03:40 | 296 |
| **London** | | | | | |
| | Apr 01 (Sat) | 13:07 | 57 | 04:37 | 306 |
| | Apr 11 (Tue) | 00:42 | 137 | 07:34 | 222 |
| | Apr 21 (Fri) | 05:21 | 62 | 21:12 | 304 |
| | Apr 30 (Sun) | 13:15 | 70 | 03:15 | 293 |

*Note that all times are in Universal Time (UT), otherwise known as Greenwich Mean Time (GMT).*

# Twilight Diagrams 2023

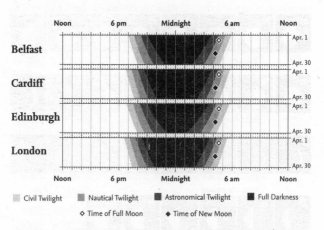

| Civil Twilight | Nautical Twilight | Astronomical Twilight | Full Darkness |

◇ Time of Full Moon    ◆ Time of New Moon

**The exact times of the Moon's major phases are shown on the diagrams opposite.**

### Sub-tropical highs

Semi-permanent areas in both hemispheres around the latitudes of approximately 30° north and south, where air that has risen at the equator descends back to the surface, becoming heated and dry as it does so. They form the descending limbs of the Hadley cells – the cells closest to the equator.

# The Moon's Phases and Ages 2023

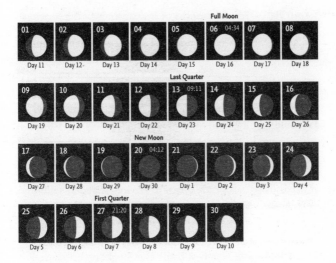

| | | | | Full Moon | | | |
|---|---|---|---|---|---|---|---|
| 01 | 02 | 03 | 04 | 05 | 06 04:34 | 07 | 08 |
| Day 11 | Day 12 | Day 13 | Day 14 | Day 15 | Day 16 | Day 17 | Day 18 |

| | | | | Last Quarter | | | |
|---|---|---|---|---|---|---|---|
| 09 | 10 | 11 | 12 | 13 09:11 | 14 | 15 | 16 |
| Day 19 | Day 20 | Day 21 | Day 22 | Day 23 | Day 24 | Day 25 | Day 26 |

| | | New Moon | | | | | |
|---|---|---|---|---|---|---|---|
| 17 | 18 | 19 | 20 04:12 | 21 | 22 | 23 | 24 |
| Day 27 | Day 28 | Day 29 | Day 30 | Day 1 | Day 2 | Day 3 | Day 4 |

| First Quarter | | | | | | | |
|---|---|---|---|---|---|---|---|
| 25 | 26 | 27 21:20 | 28 | 29 | 30 | | |
| Day 5 | Day 6 | Day 7 | Day 8 | Day 9 | Day 10 | | |

*Depression*

A low-pressure area. (Often called a 'storm' in North-American usage.) Winds circulate around a low-pressure centre in an anticlockwise direction in the northern hemisphere. (Clockwise in the southern hemisphere.) Away from the surface, and the friction that it causes, winds flow along the isobars. Depressions generally move across the globe from west to east, although under certain conditions they may linger over an area or even (rarely) move towards the west.

# April – In this month

**3 April 2001 –** Several months of heavy rainfall – the heaviest since records began in 1766 – cause some 160 metres of the chalk cliffs at Beachy Head in East Sussex to collapse into the sea.

**4 April 1635 –** Extreme hail fell at Castletown, County Offaly. The stones were 4 inches (about 100 mm) in circumference.

**5 April 1635 –** Twelve-year-old Peter Chivers invents the sport of windsurfing in a gentle breeze at Hayling Island in southern Hampshire.

**7 April 1461 –** The bloodiest battle ever fought on British soil takes place during the War of the Roses at Towton in Yorkshire. A snowstorm blinds the archers of the House of Lancaster, whose arrows fall short. It is believed that in the ensuing battle no fewer than 28,000, mainly Lancastrian soldiers, die.

**10 April 1736 –** The first swallow arrives at Stratton Strawless in Norfolk. The fact is noted down (for the first time ever) by Robert Marsham, who thus invents the science of phenology (the study of the occurrence of natural phenomena).

**21 April 1850** – Dr George Merryweather of Whitby claims that a storm off the Yorkshire coast is accurately predicted that day by his 'jury of philosophical counsellors'. His 'jury' consists of twelve leeches, and Merryweather claims that their behaviour accurately predicts the weather. In practice, of course, such claims are completely without foundation.

**23 April 1471** – The Battle of Barnet during the War of the Roses took place in such heavy mist that the soldiers of both sides could not distinguish whom they were fighting. The Earl of Warwick's men fought with those of the Earl of Oxford, thinking that they were engaging with King Edward's men. The resultant chaos marked the beginning of the end of the War of the Roses.

**25 April 1908** – A major blizzard hits the country. More than 60 cm of snow occurs in northern Hampshire and Berkshire. Oxford receives more than 42 cm of snow, and there is even about 30 cm of snow on Alderney in the Channel Islands.

# Howard

In April 1803, Luke Howard's seminal book *On the Modifications of Clouds* was published. This was initially inspired by a talk that Howard gave to the Askesian Society, a small, scientific debating society, in early 1802. Howard, by profession was a manufacturing chemist, but because of the innovative ideas put forward in his talk and the book, he has become known as 'the godfather of clouds', the 'namer of clouds' and also as 'the father of meteorology'.

The significance of this work was that Howard was the first to formalise the names given to cloud forms. Previously, clouds were regarded as too mutable to be named, but Howard recognised that certain forms were found in different types of cloud. In this, he was partly inspired by the way in which plants and animals had recently been given scientific names, thus classifying objects in natural history, but also by his extensive observations of clouds.

Howard named three principal cloud forms: cumulus, stratus and cirrus. 'Cumulus' from the Latin for 'heap'; 'stratus' from 'layer' and 'cirrus' from 'curl' or 'wisp' (of hair). He also proposed certain compound types ('modifications' as he called them) such as cirrostratus and cirrocumulus. The three principal terms and others are still used today. One term that Howard introduced, 'nimbus', meaning 'dark cloud' has now been incorporated into the term 'nimbostratus' for a specific type of cloud.

Howard's work exerted a transformation in the study of clouds and was widely recognised. John Constable, the painter, immediately started to use Howard's terminology. He made many cloud studies, calling it his 'skying' and frequently annotated his drawings and paintings with Howard's terminology. Constable's cloud studies often gave exact dates, and are indeed so detailed that they, and specifically the series made on Hampstead Heath, have been used in historical studies of the weather on particular days. Constable and other painters, such as Richard Bonnington, were particularly concerned to show realistic clouds in their paintings, so they found Howard's work of considerable significance. Howard's contribution was widely recognised. The German polymath, Johann Wolfgang von Goethe, was particularly interested in scientific subjects. Goethe was impressed with Howard's work when he read a translation in 1815, and actively promoted the use of the Latin terms for various cloud forms. He even wrote a series of poetic fragments on the various forms of cloud. This was later given as an appreciation of Howard's work, with the title *Howards Ehrengedächtnis* ('In Honour of Howard').

*An etching by Luke Howard, originally captioned 'Cumulostratus forming, fine weather cirrus above'.*

May

# Introduction

The month of May has traditionally been associated with the beginning of summer, with innumerable celebrations taking place on May 1. Many of these events are very ancient festivals to mark the first day of summer.

In England, festivities on May 1 frequently included maypole dancing (often thought to be a form of fertility rite), Morris dancing, and sometimes the appearance of a Green Man. In Scotland, May Day celebrations, which have taken place for centuries, were formerly closely associated with the ancient rite of Beltane. The latter was one of the four traditional Gaelic festivals, celebrated approximately halfway between the spring equinox and the summer solstice. Over the years it had become established as May 1. In Wales, celebrations were, like those in Scotland, originally a continuation of the ancient Beltane festivities.

Although it is true that May, more than any other month, experiences more periods of high-pressure (anticyclonic) conditions, with long spells of fine, settled weather, records show that May has also experienced its fair share of extreme weather, with major snowfalls as well as extreme heat. Bank holidays for England had been established in 1871 under William Gladstone as Prime Minister. The original intention was for a bank holiday to occur at Whitsun. But Whitsun is a religious date, seven weeks after Easter and, although it often fell in May, could fall on any day between May 11 and June 14. With such a range of dates, the weather could be extremely variable. Twenty years after the introduction of bank holidays, on the Whitsun Bank Holiday on 18 May 1891, there was a major snowfall that blanketed East Anglia. On that date, the lowest May temperature was recorded for England: -8.6°C in the notorious Rickmansworth frost hollow. Snow is not entirely uncommon this late in the year. Snow in Yorkshire was particularly deep on 17 May 1935. In mid-May 1955, the worst snowstorm for 60 years affected a large area covering Birmingham, the Cotswolds and the Chiltern Hills. On 19 May 1996, heavy snow on Dartmoor caused the annual Ten Tors adventure training weekend, run by the army, to be abandoned.

Other May weather had been very changeable. For example, that Ten Tors weekend had temperatures of 26°C in 1997, and was abandoned again in 2007, when a young girl was swept away by a brook, swollen from a depth of 1 metre to 5 metres by heavy rain. The wettest May day ever was 7 May 1881, when no less than 172.2 mm of rain fell in Cumbria. Extreme thunderstorms occurred on Derby Day, 31 May 1911. In May 1923, Scotland Yard complained that fog in London was hampering traffic and preventing the detection of crime.

---

*Synoptic*
The term 'synoptic' is used extensively in meteorology to indicate that the data used in preparing a chart (for example) were all obtained at the same time and thus show the state of the atmosphere at a particular moment.

---

# Weather Extremes

| | Country | Temp. | Location | Date |
|---|---|---|---|---|
| *Maximum temperature* | | | | |
| | England | 32.8°C | Camden Square (London) | 22 May 1922 |
| | | | Horsham (West Sussex) | 29 May 1944 |
| | | | Tunbridge Wells (Kent) | 29 May 1944 |
| | | | Regent's Park (London) | 29 May 1944 |
| | Northern Ireland | 28.0°C | Knockarevan (Co. Fermanagh) | 31 May 1997 |
| | Scotland | 30.9°C | Inverailort (Highland) | 25 May 2012 |
| | Wales | 29.2°C | Towy Castle (Carmarthenshire) | 21 May 1989 |
| *Minimum temperature* | | | | |
| | England | -9.4°C | Lynford (Norfolk) | 4 May 1941 |
| | | | | 11 May 1941 |
| | Northern Ireland | -6.5°C | Moydamlaght (Co. Londonderry) | 7 May 1982 |
| | Scotland | -7.7°C | Kinbrace (Highland) | 5 May 1981 |
| | Wales | -6.1°C | Alwen (Conwy) | 1 May 1960 |
| | | | Alwen (Conwy) | 3 May 1967 |
| | | | St Harmon (Powys) | 14 May 1984 |

| | Country | Pressure | Location | Date |
|---|---|---|---|---|
| *Maximum pressure* | | | | |
| | Eire | 1043.0 hPA | Sherkin Island (Co. Cork) | 12 May 2012 |
| | | | Valentia Obsy. (Co. Kerry) | |
| *Minimum pressure* | | | | |
| | England | 968.0 hPa | Sealand (Cheshire) | 8 May 1943 |

**Maximum temperature**
30.9°C
2012 May 25

**Minimum temperature**
-7.7°C
1981 May 05

Kinbrace

**Minimum temperature**
-6.5°C
1982 May 07

Inverailort

**Minimum temperature**
-6.1°C
1960 May 01
1967 May 03

**Minimum pressure**
968.0 hPa
1943 May 08

**Maximum temperature**
28.0°C
1997 May 31

Moydamlaght

**Minimum temperature**
-9.4°C
1941 May 04
1941 May 11

Knockarevan

Alwen

Sealand

**Maximum temperature**
32.8°C
1922 May 22

Lynford

St Harmon

Valentia Obsy.

Sherkin Island

Regent's Park

Camden Square

Towy Castle

Horsham

Tunbridge Wells

**Maximum pressure**
1043.0 hPa
2012 May 12

**Maximum temperature**
29.2°C
1989 May 21

**Maximum temperature**
32.8°C
1944 May 29

**Minimum temperature**
-6.1°C
1984 May 14

# The Weather in May 2022

| | Observation | Location | Date |
|---|---|---|---|
| **Max. temperature** | 27.51°C | Heathrow (Greater London) | 17 May |
| **Min. temperature** | -1.7°C | Altnaharra (Sutherland) | 7 May |
| **Rainfall** | 65.4 mm | Honister (Cumbria) | 10 May |
| **Wind gust** | 27 m/s (52 knots or 60 mph) | Brizlee Wood (Northumberland) | 25 & 26 May |

*Supercell*
A supercell is an extremely violent, persistent thunderstorm that is marked by an extremely large, rotating updraught or 'mesocyclone'. Supercells are accompanied by heavy rain, large hail and frequent cloud-to-ground lightning discharges. The updraught may extend as high as 15 km into the atmosphere. The updraught is accompanied by strong downdraughts, but the two streams of air are separated horizontally in space, and this is one reason for a supercell's long lifetime (sometimes many hours) when compared with 'ordinary' thunderstorms, which may persist for about one hour. Cool downdraught air often bleeds into the mesocyclone and is the site of the formation of tornadoes.

In general, the weather in May 2022 was quiet, with some scattered showers and more prolonged periods of rain. Weather systems that arrived from the Atlantic produced near-average amounts of rain over much of the country. The western part of Scotland was noticeably wet, with double the expected average rainfall in a few locations. Northern Ireland was also rather wetter than expected.

In the middle of the month, high pressure, which had been affecting the country, slowly drifted away towards the Continent and a plume of warm, unstable air moved across the Channel from France. This led to the formation of heavy showers and even thunderstorms in various parts of the country (including Northern Ireland) between May 16 and 19. There were heavy downpours in Northern Ireland on May 16, with localised flooding in the Belfast area and in Derry. Once again, Honister Pass in Cumbria recorded the most rainfall in 24 hours, with 65.4 mm on May 10. South-east England experienced widespread thunderstorms during the night of May 18/19, with extensive lightning activity and some resulting damage to properties. Some parts of the counties along the South Coast were particularly dry, with little or no rain during the month.

The warm air incursion in the middle of the month led to relatively high temperatures over much of the country, in addition to breeding heavy showers and thunderstorms, with the month's maximum of 27.5°C recorded at Heathrow (Greater London) on May 17. The minimum temperature of -1.7°C for the month was observed at Altnaharra in Sutherland, more than a week earlier, on May 7.

The end of the month was generally cooler, but still with some heavy showers and thunderstorm activity.

# Sunrise and Sunset 2023

| Location | Date | Rise | Azimuth ° | Set | Azimuth ° |
|----------|------|------|-----------|-----|-----------|
| **Belfast** | | | | | |
| | May 01 (Mon) | 04:46 | 62 | 19:57 | 298 |
| | May 11 (Thu) | 04:26 | 57 | 20:16 | 304 |
| | May 21 (Sun) | 04:09 | 52 | 20:33 | 308 |
| | May 31 (Wed) | 03:56 | 48 | 20:47 | 312 |
| **Cardiff** | | | | | |
| | May 01 (Mon) | 04:45 | 64 | 19:35 | 296 |
| | May 11 (Thu) | 04:28 | 59 | 19:52 | 301 |
| | May 21 (Sun) | 04:13 | 55 | 20:06 | 305 |
| | May 31 (Wed) | 04:03 | 52 | 20:19 | 308 |
| **Edinburgh** | | | | | |
| | May 01 (Mon) | 04:30 | 61 | 19:51 | 299 |
| | May 11 (Thu) | 04:09 | 55 | 20:11 | 305 |
| | May 21 (Sun) | 03:51 | 51 | 20:29 | 310 |
| | May 31 (Wed) | 03:37 | 47 | 20:45 | 314 |
| **London** | | | | | |
| | May 01 (Mon) | 04:33 | 64 | 19:24 | 296 |
| | May 11 (Thu) | 04:16 | 59 | 19:40 | 301 |
| | May 21 (Sun) | 04:01 | 55 | 19:55 | 305 |
| | May 31 (Wed) | 03:50 | 52 | 20:08 | 308 |

*Note that all times are in Universal Time (UT), otherwise known as Greenwich Mean Time (GMT). These times do not take Summer Time (BST) into account.*

# Moonrise and Moonset 2023

| Location | Date | Rise | Azimuth ° | Set | Azimuth ° |
|----------|------|------|-----------|-----|-----------|
| **Belfast** | | | | | |
| | May 01 (Mon) | 14:48 | 79 | 03:55 | 285 |
| | May 11 (Thu) | 02:11 | 138 | 09:00 | 224 |
| | May 21 (Sun) | 04:39 | 40 | 23:26 | 323 |
| | May 31 (Wed) | 16:20 | 107 | 02:26 | 259 |
| **Cardiff** | | | | | |
| | May 01 (Mon) | 14:40 | 80 | 03:39 | 284 |
| | May 11 (Thu) | 01:38 | 134 | 09:09 | 228 |
| | May 21 (Sun) | 04:51 | 44 | 22:48 | 318 |
| | May 31 (Wed) | 16:02 | 105 | 02:19 | 260 |
| **Edinburgh** | | | | | |
| | May 01 (Mon) | 14:34 | 79 | 03:47 | 286 |
| | May 11 (Thu) | 02:11 | 140 | 08:38 | 222 |
| | May 21 (Sun) | 04:16 | 37 | 23:28 | 326 |
| | May 31 (Wed) | 16:11 | 197 | 02:13 | 259 |
| **London** | | | | | |
| | May 01 (Mon) | 14:28 | 80 | 03:27 | 284 |
| | May 11 (Thu) | 01:27 | 134 | 08:56 | 228 |
| | May 21 (Sun) | 04:38 | 44 | 22:37 | 318 |
| | May 31 (Wed) | 15:50 | 105 | 02:07 | 260 |

*Note that all times are in Universal Time (UT), otherwise known as Greenwich Mean Time (GMT). These times do not take Summer Time (BST) into account.*

# Twilight Diagrams 2023

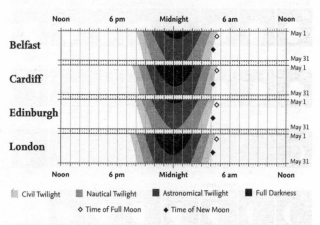

| | Civil Twilight | | Nautical Twilight | | Astronomical Twilight | | Full Darkness |
|---|---|---|---|---|---|---|---|

◇ Time of Full Moon ◆ Time of New Moon

The exact times of the Moon's major phases are shown on the diagrams opposite.

---

### *Adiabatic*

Any process in which heat does not enter or leave the system. Air rising in the troposphere generally cools at an adiabatic rate, because it does not lose heat to its surroundings. The fall in temperature is solely because of its expansion: its increase in volume, because of the decrease in pressure with increasing altitude.

---

# The Moon's Phases and Ages 2023

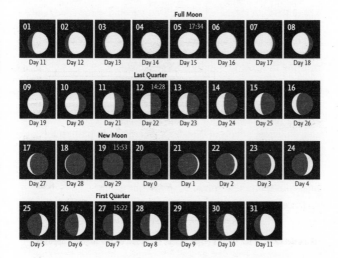

**Full Moon**

| | | | | | | | |
|---|---|---|---|---|---|---|---|
| 01 | 02 | 03 | 04 | 05 17:34 | 06 | 07 | 08 |
| Day 11 | Day 12 | Day 13 | Day 14 | Day 15 | Day 16 | Day 17 | Day 18 |

**Last Quarter**

| 09 | 10 | 11 | 12 14:28 | 13 | 14 | 15 | 16 |
|---|---|---|---|---|---|---|---|
| Day 19 | Day 20 | Day 21 | Day 22 | Day 23 | Day 24 | Day 25 | Day 26 |

**New Moon**

| 17 | 18 | 19 15:53 | 20 | 21 | 22 | 23 | 24 |
|---|---|---|---|---|---|---|---|
| Day 27 | Day 28 | Day 29 | Day 0 | Day 1 | Day 2 | Day 3 | Day 4 |

**First Quarter**

| 25 | 26 | 27 15:22 | 28 | 29 | 30 | 31 |
|---|---|---|---|---|---|---|
| Day 5 | Day 6 | Day 7 | Day 8 | Day 9 | Day 10 | Day 11 |

*Hurricane*
The term used for a tropical cyclone in the North Atlantic Ocean or eastern Pacific Ocean. Hurricanes (indeed all tropical cyclones) are driven by high sea-surface temperatures, and cannot occur over the British Isles.

# May – In this month

**6 May 1954 –** A strong wind was blowing all day and it was only late in the afternoon when it had started to moderate slightly and blow in gusts that Roger Bannister decided to attempt to run one mile in less than four minutes. He succeeded in running the mile in 3 minutes 59.4 seconds.

**11 May 1968 –** The Rugby League Final, between Leeds and Wakefield Trinity at Wembley, took place, in part, in a thunderstorm. The pitch was completely flooded and the players found it almost impossible to keep their footing. The 'man of the match', Don Fox, the Wakefield prop, missed the final conversion that would have won the game.

**11 May 1785 –** Gilbert White, the naturalist and author, wrote in his diary for this day, describing conditions at his home in Selbourne in Hampshire, 'Cloudless days. The country all dust.'

**12 May 1315 –** This day saw the beginning of a sustained spell of bad weather. The rain began and continued for so long that seed failed to germinate, being washed away. There were extremely extensive floods and many flocks of sheep were drowned. The 'Great Famine' that this flooding precipitated lasted in England from 1315 to 1318.

**28 May 1821 –** There is an unprecedented late heavy snowfall in north Yorkshire in both the town of Halifax and in the city of Leeds.

**31 May 2022 –** A 3.8-magnitude earthquake was recorded in the centre of Shropshire. It was the third earthquake to occur in the United Kingdom within 24 hours.

**May 1975 –** This month saw the highest monthly amount of sunshine ever recorded anywhere in Scotland. Tiree in Argyll and Bute experienced 329.1 hours of sunshine.

*Humidity*
A measure of the quantity of water vapour in the air. It generally increases with an increase in temperature.

# Tornadoes

Tornadoes are often reported in Britain, indeed it has been stated that more tornadoes are seen in Britain than in any other country in the world. This is probably not a result of a greater frequency of events, but more probably down to a greater population density. Although tornadoes in the United States are undoubtedly larger and more dangerous, if one occurs over uninhabited farmland in one of the Prairie States, it may well go unreported until crop damage is noticed. The way in which tornadoes are formed must also play a part. The strongest and most dangerous tornadoes form from supercell storms, where there is what is known as a mesocyclone, a large single, rotating storm. Supercells are rare in Britain, although they do occur.

Many 'tornadoes' or 'twisters' reported in the media are probably not tornadoes as such, but 'landspouts', which form by a completely different mechanism, or even 'gustnadoes', which are extremely strong whirls that may occur on powerful gust fronts.

Landspouts (a term not often used) and waterspouts form beneath powerful cumuliform clouds (cumulus, cumulus congestus or cumulonimbus) when a powerful downdraught reaches the surface. True tornadoes form initially as a rotating horizontal roll, one end of which is lifted by the extreme up- and down-draughts in a large rotating supercell.

The longest-lived tornado in Britain (and one of the longest in Europe) occurred on 21 May 1950, and tracked for at least 107.1 km. Depending on the account given, it was first seen either at Little London in Buckinghamshire, a few kilometres east of Oxford, or farther east at Puttenham in Berkshire. It tracked as far as Coveney in Cambridgeshire. There it lifted from the surface and became a funnel cloud, and travelled another 52.6 km to Shipham in Norfolk, after which it was seen disappearing out into the North Sea. There were reports of damage all the way along its surface track, so it is assumed to have been a single tornado, rather than a series of individual tornadoes.

The intensity of tornadoes in the United Kingdom is normally expressed on the TORRO scale, which was developed by TORRO (The Tornado and Storm Research Organisation). This is described in more detail on page 252. It is based on wind speed, and thus differs from the Enhanced Fujita Scale, used in the United States, which is based on the damage caused by any particular tornado.

---

*Rain shadow*
An area to the leeward of high ground, whether hills or mountains, often experiences less rainfall than neighbouring areas or than expected. This rain-shadow effect occurs because as air rises over the higher ground (usually to the west) there is increased rainfall, leaving less moisture to fall on any areas to the leeward of the hills.

---

June

# Introduction

Although meteorologists regard the month of June as marking the start of the summer season, and it may see the longest day at the summer solstice (June 21), in Britain it includes the hottest day just about one year in four. This is about the same as the frequency in the month of August. The hottest day is most common in July. To the general public, June may have come to be associated with the catchphrase 'flaming June', but it is not often accompanied by particularly hot weather, and is only very rarely warmer than July. It shows little sign of getting warmer over time. During the last three centuries, the month has been about as warm as September.

In the middle of the month, the weather (when regarded as showing five seasons), tends to change quite suddenly from 'spring and early summer' to 'high summer'. This sees a quite sudden resumption of predominant westerly conditions from the mixed regimes that prevailed earlier in the year. However, the reduced temperature contrast at this time of year over the Atlantic results in weaker winds and the depressions that arrive from the Atlantic are slow-moving, so any accompanying rain tends to linger and be slow to move away. This striking change in the prevailing weather is often considered to be the start of the 'European monsoon', marked by slow-moving depressions that cross the British Isles and track towards the Baltic or farther northward towards Scandinavia.

*High summer – June 18 to September 9*
There is a significant change in the overall circulation in early June, from the changeable situation that has prevailed over the preceding three months, when northerly and easterly airflow often predominated. In mid-June, the dominant westerly circulation is restored, with persistent westerly and north-westerly winds and their accompanying depressions. More settled conditions sometimes arise, when the Azores High extends a ridge of high pressure towards western Europe and occasionally over the British Isles, resulting in a fine, warm, dry summer.

*Previous page:* Flaming June, *the most famous painting (often regarded as his masterpiece) by Frederic Leighton, painted in 1895.*

# Weather Extremes

|  | Country | Temp. | Location | Date |
|---|---|---|---|---|
| **Maximum temperature** | | | | |
|  | England | 35.6°C | Mayflower Park, Southampton (Hampshire) | 28 Jun. 1976 |
|  | Northern Ireland | 30.8°C | Knockaraven (Co. Fermanagh) | 30 Jun. 1976 |
|  | Scotland | 32.2°C | Ochtertyre (Perth & Kinross) | 18 Jun. 1893 |
|  | Wales | 33.5°C | Usk (Monmouthshire) | 28 Jun. 1976 |
| **Minimum temperature** | | | | |
|  | England | -5.6°C | Santon Downham (Norfolk) | 1 Jun. 1962 3 Jun. 1962 |
|  | Northern Ireland | -2.4°C | Lough Navar Forest (Co. Fermanagh) | 4 Jun. 1991 |
|  | Scotland | -5.6°C | Dalwhinnie (Inverness-shire) | 9 Jun. 1955 |
|  | Wales | -4.0°C | St Harmon (Powys) | 8 Jun. 1985 |

|  | Country | Pressure | Location | Date |
|---|---|---|---|---|
| **Maximum pressure** | | | | |
|  | Eire | 1043.1 hPa | Clones (Co. Monaghan) | 14 Jun. 1959 |
| **Minimum pressure** | | | | |
|  | Scotland | 968.4 hPa | Lerwick (Shetland) | 28 Jun. 1938 |

**Minimum pressure**
968.4 hPa
1938 Jun 28
Lerwick

**Maximum temperature**
32.2°C
1893 Jun 18

**Minimum temperature**
-5.6°C
1955 Jun 09

**Maximum pressure**
1043.1 hPa
1959 Jun 14

Dalwhinnie

Ochtertyre

**Minimum temperature**
-2.4°C
1991 Jun 04

Lough Navar Forest    Clones
Knockareven

**Minimum temperature**
-5.6°C
1962 Jun 01
1962 Jun 03

**Maximum temperature**
30.8°C
1976 Jun 30

Stanton Downham

St Harmon

Usk

**Minimum temperature**
-4.0°C
1985 Jun 08

Mayflower Park, Southampton

**Maximum temperature**
33.5°C
1976 Jun 28

**Maximum temperature**
35.6°C
1976 Jun 28

# The Weather in June 2021

| Observation | Location | Date |
| --- | --- | --- |
| *Max. temperature*<br>29.7°C | Teddington Bushy Park<br>(Greater London) | 14 June |
| *Min. temperature*<br>-2.4°C | Altnaharra<br>(Sutherland) | 22 June |
| *Rainfall*<br>74.0 mm | Princetown (Devon) | 28 June |
| *Wind gust*<br>24 m/s<br>(47 knots or 54 mph) | Loch Gascarnoch<br>(Ross & Cromarty) | 10 June |

The month of June 2021 was fairly quiet, dry and warm, although it was less settled over Scotland in the middle of the month. There were some heavy showers and thunderstorms, particularly in the south-east of England. The south-east was also extremely wet, with more than double the normal amount of rain in some locations. Generally, all areas from the Midlands northwards had a dry month with less rain than normal.

The month started dry in England, although rain in the form of heavy showers (and even some thunderstorms) spread across the country from the south-west. The month's highest temperature of 29.7°C was recorded at Teddington Bushy Park on June 14. In south-eastern and eastern England there was extensive flooding on June 17 and 18 from extremely heavy showers and thunderstorms as a frontal boundary became very active. Flooding extended from northern Hampshire and parts of Buckinghamshire right across the region as far as Lowestoft in Norfolk. There were also reports of flooding from torrential rain the same day in eastern Dorset.

Later in the month (June 24), patchy rain developed into more persistent and heavier showers as it moved south, with actual thunderstorms developing over Teeside and the North York Moors. The next day (June 25) a number of thunderstorms developed in northern Wales and the Midlands. There was some flooding around Birmingham. Farther south and east, there was also flooding in the London area and in Essex. A confirmed tornado occurred in Barking, East London, causing damage to roads and houses and a suspected second in the same general area also caused some damage to cars and fences. Three days later (June 28) there was more flooding in the north London area.

The very end of the month (June 28 and 29) saw heavy, persistent rain in central and southern counties of England and some thunderstorms on the south coast. Maximum rainfall of 74.0 mm for the month occurred at Princetown on Dartmoor in Devon on June 28. Persistent, heavy downpours affected areas around the Solent, where as much as 50 mm was recorded in just 12 hours.

Wales was notably dry during June 2021, although scattered showers and longer periods of rain affected most parts at some time during the month. Northern Ireland too was generally warm and dry, with just a few wetter days.

The month started very mild in northern Scotland with the temperature at Achnagart in Ross and Cromarty not dropping below 16.2°C during the night of June 1. By contrast, a minimum temperature of -2.4°C was recorded at Altnaharra in Sutherland on June 22, an extremely low temperature for the time of year.

# Sunrise and Sunset 2023

| Location | Date | Rise | Azimuth ° | Set | Azimuth ° |
|---|---|---|---|---|---|
| **Belfast** | | | | | |
| | Jun 01 (Thu) | 03:55 | 48 | 20:49 | 312 |
| | Jun 11 (Sun) | 03:48 | 46 | 20:59 | 314 |
| | Jun 21 (Wed) | 03:47 | 45 | 21:04 | 315 |
| | Jun 30 (Fri) | 03:51 | 46 | 21:03 | 314 |
| **Cardiff** | | | | | |
| | Jun 01 (Thu) | 04:02 | 52 | 20:20 | 308 |
| | Jun 11 (Sun) | 03:56 | 50 | 20:29 | 310 |
| | Jun 21 (Wed) | 03:55 | 49 | 20:34 | 311 |
| | Jun 30 (Fri) | 03:59 | 49 | 20:33 | 310 |
| **Edinburgh** | | | | | |
| | Jun 01 (Thu) | 03:36 | 46 | 20:46 | 314 |
| | Jun 11 (Sun) | 03:28 | 44 | 20:57 | 316 |
| | Jun 21 (Wed) | 03:26 | 43 | 21:03 | 317 |
| | Jun 30 (Fri) | 03:31 | 44 | 21:02 | 316 |
| **London** | | | | | |
| | Jun 01 (Thu) | 03:49 | 52 | 20:09 | 309 |
| | Jun 11 (Sun) | 03:44 | 50 | 20:18 | 311 |
| | Jun 21 (Wed) | 03:43 | 49 | 20:23 | 311 |
| | Jun 30 (Fri) | 03:47 | 49 | 20:22 | 311 |

*Note that all times are in Universal Time (UT), otherwise known as Greenwich Mean Time (GMT). These times do not take Summer Time (BST) into account.*

# Moonrise and Moonset 2023

| Location | Date | Rise | Azimuth ° | Set | Azimuth ° |
|---|---|---|---|---|---|
| **Belfast** | | | | | |
| | Jun 01 (Thu) | 17:45 | 117 | 02:35 | 249 |
| | Jun 11 (Sun) | 01:21 | 100 | 12:53 | 265 |
| | Jun 21 (Wed) | 06:23 | 45 | 23:47 | 320 |
| | Jun 30 (Fri) | 18:15 | 133 | 01:05 | 234 |
| **Cardiff** | | | | | |
| | Jun 01 (Thu) | 17:22 | 115 | 02:31 | 251 |
| | Jun 11 (Sun) | 01:06 | 100 | 12:43 | 265 |
| | Jun 21 (Wed) | 06:31 | 49 | 23:19 | 306 |
| | Jun 30 (Fri) | 17:44 | 129 | 01:08 | 237 |
| **Edinburgh** | | | | | |
| | Jun 01 (Thu) | 17:38 | 118 | 02:20 | 248 |
| | Jun 11 (Sun) | 01:11 | 101 | 12:40 | 265 |
| | Jun 21 (Wed) | 06:01 | 43 | 23:44 | 311 |
| | Jun 30 (Fri) | 18:12 | 135 | 00:47 | 232 |
| **London** | | | | | |
| | Jun 01 (Thu) | 17:10 | 115 | 02:19 | 251 |
| | Jun 11 (Sun) | 00:54 | 100 | 12:30 | 265 |
| | Jun 21 (Wed) | 06:18 | 49 | 23:08 | 307 |
| | Jun 30 (Fri) | 17:32 | 129 | 00:56 | 236 |

*Note that all times are in Universal Time (UT), otherwise known as Greenwich Mean Time (GMT). These times do not take Summer Time (BST) into account.*

# Twilight Diagrams 2023

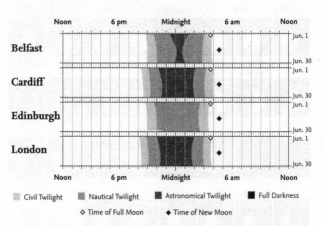

| | | | | |
|---|---|---|---|---|
| Civil Twilight | Nautical Twilight | Astronomical Twilight | Full Darkness |
| ◇ Time of Full Moon | | ◆ Time of New Moon | |

**The exact times of the Moon's major phases are shown on the diagrams opposite.**

*Noctilucent clouds*

Clouds seen in midsummer in the middle of the night (the name means 'night-shining') and (for Britain) in the general direction of the North Pole. These clouds (NLC) are seen when the observer is in darkness, but the clouds – which are the highest in the atmosphere at about 185 kilometres, far above all other clouds – are still illuminated by the Sun, which is below the horizon. They consist of ice crystals, believed to form around meteoritic dust arriving from space. They also occur in the southern hemisphere, although because of the distribution of land masses, are commonly seen only by observers in the Antarctic Peninsula (see pages 110 and 111).

# The Moon's Phases and Ages 2023

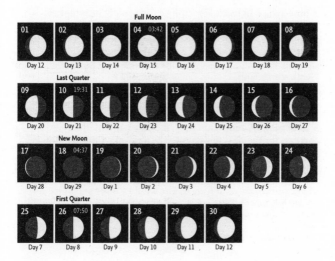

---

*Aurora*

A luminous event occurring in the upper atmosphere between approximately 100 and 1000 km. It arises when energetic particles from the Sun raise atoms to higher energy levels. When the atoms drop back to their original energy level, they emit the characteristic green and red shades that are visible to human eyes (from oxygen and nitrogen, respectively).

---

# June – In this month

**2 June 1994** – In thick fog, an RAF Chinook helicopter flew into a hillside on the Mull of Kintyre. The crash killed all four crew and 25 passengers, consisting of practically all the senior Northern Irish members of military intelligence (MI5). The disaster was a major blow to efforts to combat IRA terrorism in Northern Ireland.

**4 June 1944** – Group Captain J.M. Stagg has to deliver the forecast for the next few days to the Supreme Commander of the Allied Expeditionary Force, about to invade northern France in Operation Overlord. The unfavourable forecast for conditions in the English Channel leads to the postponement of the operation until June 6. The window of opportunity is very small, but Stagg's forecast is subsequently proved correct and the D-Day operation is ultimately successful.

**5 June 1916** – Lord Kitchener, Secretary of State for War is killed when, as a result of a severe (Force 9) storm off the Orkneys, the ship on which he is travelling to Murmansk in Russia, *HMS Hampshire*, strikes a mine and sinks with the loss of 643 lives, including that of Lord Kitchener.

**10 June 1936** – John Hassall, the poster artist, famous for the creation of the poster of the 'Jolly Fisherman' with the slogan 'Skegness is so bracing', visits the town for the first and only time, on a day of bright sunshine. He receives 'the freedom of the seafront' for the part he has played in promoting the town, 28 years after he created the poster.

**11 June 1963** – More than 80 mm of rain falls on Dublin in one hour, out of a total of 184 mm for the day. Statistically, Dublin can expect no more than 35 mm once in 50 years.

**11 June 1963** – Pershore, in Worcestershire, experiences a downpour in a storm on the same day of the year as the 'Dublin Deluge'. A total of 67 mm of rain is recorded in just 25 minutes.

**16 June 1904** – This day is 'Bloomsday', the day in which James Joyce sets his novel Ulysses in Dublin. In contrast to the warm sunny weather in the novel, the actual weather is very prosaic, with a Force 4 south-westerly wind, stratocumulus cloud, some sunshine, and rain later.

**18 June 1764** – The steeple of the ancient church of St Bride's in Fleet Street, London is struck by lightning and destroyed. The first church here was erected in the sixth century. The same storm sees the collapse of Lot's Wife, a tall spire of chalk that inspired the name given to the Needles at the western end of the Isle of Wight.

**23 June 1858** – Today sees the beginning of 'The Great Stink'. There is a heatwave in London, and the temperature of the Thames reaches 70°F (21°C). The Thames basically becomes a river of raw sewage. A bill for purification of the Thames is rapidly passed by Parliament, and the next year, Joseph Bazalgette begins his grand scheme for an effective drainage system for London.

**24 June 1967** – A caving expedition of ten pot-holers, mainly from Leeds University, goes dramatically wrong in Wharfedale, North Yorkshire. After a long spell of dry weather, torrential rain suddenly floods the caverns. Four of the team leave early, but the others remain. Eventually all six are drowned, in what becomes known as the Mossdale Caving Disaster.

# The Great Stink

Occasionally, weather events precipitate major change in urban infrastructure. This was the case with the heatwave that affected London in June, July and August 1858. The River Thames had become, in effect, an open sewer. The population had increased greatly (doubled) in the decades since 1800, and a commission in 1847 had abolished the use of cesspits. As a result all sewage was discharged, untreated, into the river, where it merely washed up and down with the tide. (The Thames is tidal as far upstream as Teddington, many miles west of the capital.) The temperature of the river itself rose so high that the water was evaporating into the air. This produced an overpowering odour, which became known as 'The Great Stink'. The problem forced itself on the members of the Houses of Parliament, which, of course, lie alongside the river.

The smell was so bad that an unsuccessful attempt was made to lessen the offensive odour by soaking canvas in chloride of lime and hanging this over the windows of the Houses of Parliament. Because of the severity of the situation, and because it was of immediate concern to Members of Parliament, a Bill was passed within a month for the purification of the River Thames. The engineer entrusted with this work, Joseph Bazalgette, saw his proposal accepted, and work began the next year, 1859, on his scheme for three interconnected high-, middle- and low-level sewers on each side of the river that would intercept all the waste and transfer everything to treatment works far to the east of the city. Most of the flow was controlled by gravity, but pumping stations were installed to lift sewage from low-level to higher sewers. Two of the stations (at Abbey Mills in Stratford and at Crossness on the Erith Marches) are considered to be of considerable architectural merit and are listed for protection by Historic England. Work on Bazalgette's scheme was finished in 1875.

The sets of intercepting sewers remain in use today, even though the population has increased dramatically from some hundreds of thousands of inhabitants to millions. Under Bazalgettte's scheme three embankments were built alongside the Thames. These are the Victoria, Chelsea and Albert Embankments.

*A drawing from the magazine* Punch *for 1858, with the caption 'The silent highwayman'. Death rows on the Thames, claiming the lives of victims who have not paid to have the river cleaned up, during the Great Stink.*

# Noctilucent Clouds

For about one month on either side of midsummer, it is sometimes possible at night (even in the middle of the night) to see glowing clouds when looking in the direction of the pole. These are very high clouds, known as noctilucent clouds. (The name means 'night-shining'.) From Britain they are seen in the general direction of the North Pole and tend to be mostly observed from Scotland, although on occasions there are major displays that may be seen from anywhere in the country.

The period of NLC visibility is actually about one month before the summer solstice until roughly two months after the solstice. It is believed that this inequality is caused by the time for the atmosphere to cool sufficiently for the necessary temperatures to be reached at that altitude. The 'spike' in activity in 2022 may be related to water vapour introduced by the SpaceX launch of the Globalstar satellite on June 19.

NLC are seen from both hemispheres, although less frequently reported from the south, probably because of the low population density and the small amount of land from which they may be observed.

These clouds (NLC) are seen when the observer is in darkness, but the clouds – which are the highest in the atmosphere at about 185 kilometres, far above all other clouds – are still illuminated by the Sun, which is below the northern horizon. (The 'normal' clouds, such as those described on pages 33–42 are found in the lowest layer of the atmosphere, the troposphere, which, even in the tropics, extends to an altitude of no more than 20 km.)

Noctilucent clouds consist of ice crystals, believed to form around particles of meteoritic dust arriving from space. Although they are known to be formed of ice, the origin of the water that freezes into ice crystals is still unknown and the subject of great debate. For many years it was believed that the water also came 'from outside' and was brought by comets and other bodies. There is another possibility. Although water vapour cannot be carried to such great heights, it is possible that water vapour may be created by the breakdown of methane gas, which can rise freely to such extreme altitudes and then be broken down by radiation from the Sun.

*Noctilucent clouds, photographed by Alan Tough from Nairn in Scotland, on 31 May 2020, at 00:28.*

July

# Introduction

The saying that 'the English summer consists of three fine days and a thunderstorm' has been ascribed to both the kings Charles II and George III. However, it is probably a proverbial piece of weather lore, the origins of which are lost. A succession of hot days and humid air certainly provides the conditions for the formation of giant cumulonimbus clouds and thunderstorms with the accompanying torrential rain or even hail. Such a situation usually ends a heatwave (at least for a day or so) and is very typical of British weather.

July is definitely associated with high summer, and it is usually the hottest month of the year. It frequently includes the hottest day of the year. This occurs about 44 per cent of the time, the remaining percentage of hottest days being more-or-less equally divided between June and August. In 1976 it definitely included the hottest day, when the temperature reached 35.9°C at Cheltenham on 3 July 1976.

The year 1976 was the 'drought year', when, beginning in late June, there were ten weeks of sunshine and practically no rain. (It was actually the second driest summer of the twentieth century, after 1995.) A very few locations were fortunate and had some rain, although this was generally less than half of the usual rainfall for the month of July. A high-pressure 'block' over the UK diverted the normal succession of depressions from the Atlantic, and their accompanying rain, south towards the Mediterranean. The drought only came to an end (in late August) just after the government appointed a Drought Minister.

The prolonged drought was actually the result of a very long period of reduced rainfall. In the preceding year, 1975, both the summer and autumn were dry, the winter 1975–76 was particularly dry and then so was the spring of 1976. (Here we use the three-month 'meteorological' seasons, rather than the five seasons that we discuss elsewhere, page 11.) Certain areas of the country actually experienced months with no rain. Portions of south-west England had no rain at all in July and in the first half of August 1976.

July 2006 was the warmest calendar month ever recorded in the Central England Temperature series, which began in 1659. There was an unusual high pressure region over northern Europe and a persistent airstream from the south affecting the British Isles.

---

### Icelandic Low

A semi-permanent feature of the distribution of pressure over the North Atlantic. Unlike the more-or-less permanent Azores High, it largely arises because depressions (low-pressure systems) frequently pass across the area. A similar low-pressure area exists over the northern Pacific Ocean, often known as the 'Aleutian Low'.

---

# Weather Extremes

| Country | Temp. | Location | Date |
|---|---|---|---|
| **Maximum temperature** | | | |
| England | 37.8°C | Heathrow (London) | 31 Jul. 2020 |
| Northern Ireland | 31.3°C | Castlederg (Co. Tyrone) | 21 Jul. 2021 |
| Scotland | 32.8°C | Dumfries (Dumfries & Galloway) | 20 Jul. 1901 / 2 Jul. 1908 |
| Wales | 34.6°C | Gogerddan (Ceredigion) | 19 Jul. 2006 |
| **Minimum temperature** | | | |
| England | -1.7°C | Kielder Castle (Northumberland) | 17 Jul. 1965 |
| Northern Ireland | -1.1°C | Lislap Forest (Co. Tyrone) | 17 Jul. 1971 |
| Scotland | -2.5°C | Lagganlia (Inverness-shire) | 15 Jul. 1977 |
| Wales | -2.5°C | St Harmon (Powys) | 9 Jul. 1986 |

| Country | Pressure | Location | Date |
|---|---|---|---|
| **Maximum pressure** | | | |
| Scotland | 1039.2 hPa | Aboyne (Aberdeenshire) | 16 Jul. 1996 |
| **Minimum pressure** | | | |
| Scotland | 967.9 hPa | Sule Skerry (Northern Isles) | 8 Jul. 1964 |

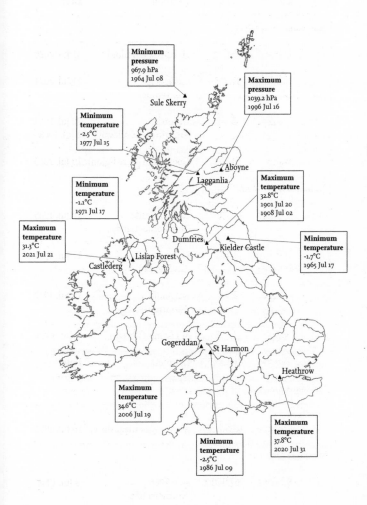

**Minimum pressure**
967.9 hPa
1964 Jul 08

Sule Skerry

**Maximum pressure**
1039.2 hPa
1996 Jul 16

**Minimum temperature**
-2.5°C
1977 Jul 15

Aboyne

Lagganlia

**Maximum temperature**
32.8°C
1901 Jul 20
1908 Jul 02

**Minimum temperature**
-1.1°C
1971 Jul 17

Dumfries

Kielder Castle

**Minimum temperature**
-1.7°C
1965 Jul 17

**Maximum temperature**
31.3°C
2021 Jul 21

Lislap Forest

Castlederg

Gogerddan

St Harmon

Heathrow

**Maximum temperature**
34.6°C
2006 Jul 19

**Minimum temperature**
-2.5°C
1986 Jul 09

**Maximum temperature**
37.8°C
2020 Jul 31

# The Weather in July 2021

| Observation | Location | Date |
| --- | --- | --- |
| *Max. temperature* | | |
| 32.2°C | Heathrow (Greater London) | 20 July |
| 31.3°C | Castlederg (County Tyrone) (Northern Ireland record) | 21 July |
| *Min. temperature* | | |
| -0.1°C | Braemar (Aberdeenshire) | 2 July |
| *Rainfall* | | |
| 87.9 mm | Bethersden (Kent) | 26 July |
| *Wind gust* | | |
| 34.5 m/s (67 knots or 77 mph) | Needles Old Battery (Isle of Wight) | 30 July |

July 2021 was rather similar to June that year, with extremely unsettled weather in the south-east of England. Some areas around London experienced extremely heavy falls of rain – which combined to give some areas more than four times the normal amount of rainfall – whereas neighbouring areas (such as north Kent) had about half the usual amount.

The first twelve days of July 2021 were generally unsettled, with spells of heavy rain and showers, particularly over English counties, while other parts of Britain had less rain and greater warmth. The whole of the United Kingdom was drier and warmer by mid-month particularly in the south-west and the north of Scotland, as well as in Northern Ireland. Temperatures exceeded 30°C on several days in some areas, with unbroken sunshine. In Northern Ireland, 31.3°C was recorded on the 21st at Castlederg, County Tyrone, and this set a new record for the region as the highest temperature recorded in any month. The final week was much more unsettled everywhere, with lower temperatures and frequent showery rain. Many parts of England, and even portions of the Scottish Highlands, had an extremely wet month, with more than double the usual amount of rainfall.

The unsettled weather early in the month brought heavy showers to many areas. There was flooding in both the Edinburgh and Glasgow areas of Scotland, around Manchester, in Northern Ireland, as well as in the south-west of England, with both road and rail disruptions. Flooding affected eastern England, with evacuations in Peterborough and traffic disruption farther east in Norwich. Flooding and high winds also brought road and rail problems to the south-east, with severe conditions in the London area and also delays to many ferry crossings of the English Channel. Flooding was reported in many southern counties as far west as Dorset.

Towards the end of the month (from July 20 onwards) renewed heavy showers and thunderstorms affected many parts of England. There was damaging large hail in Leicestershire and flooding in the area with road closures in both Leicestershire and Nottinghamshire. Lightning caused an interruption to power supplies in Hertfordshire and other problems in the south-east. There was rail disruption in Kent because of flooding.

At the very end of July, thunderstorms caused problems in Scotland and many southern and Midland counties. On July 30, the south was hit by Storm Evert, which brought high winds to the area. The month's maximum wind gust was recorded at the Needles Old Battery on the Isle of Wight. There were power outages, and fallen trees blocked many roads, with Isle of Wight hovercraft services cancelled. Flooding affected areas of Leicestershire and Derbyshire.

# Sunrise and Sunset 2023

| Location | Date | Rise | Azimuth ° | Set | Azimuth ° |
|----------|------|------|---------|-----|---------|
| **Belfast** | | | | | |
| | Jul 01 (Sat) | 03:52 | 46 | 21:03 | 314 |
| | Jul 11 (Tue) | 04:02 | 46 | 20:56 | 312 |
| | Jul 21 (Fri) | 04:16 | 51 | 20:44 | 308 |
| | Jul 31 (Mon) | 04:32 | 56 | 20:27 | 304 |
| **Cardiff** | | | | | |
| | Jul 01 (Sat) | 04:00 | 50 | 20:33 | 310 |
| | Jul 11 (Tue) | 04:08 | 51 | 20:27 | 308 |
| | Jul 21 (Fri) | 04:20 | 54 | 20:17 | 305 |
| | Jul 31 (Mon) | 04:35 | 58 | 20:03 | 301 |
| **Edinburgh** | | | | | |
| | Jul 01 (Sat) | 03:32 | 44 | 21:01 | 316 |
| | Jul 11 (Tue) | 03:42 | 46 | 20:53 | 314 |
| | Jul 21 (Fri) | 03:57 | 50 | 20:40 | 310 |
| | Jul 31 (Mon) | 04:15 | 54 | 20:22 | 305 |
| **London** | | | | | |
| | Jul 01 (Sat) | 03:47 | 49 | 20:22 | 311 |
| | Jul 11 (Tue) | 03:56 | 52 | 20:16 | 309 |
| | Jul 21 (Fri) | 04:08 | 54 | 20:06 | 305 |
| | Jul 31 (Mon) | 04:22 | 58 | 19:52 | 301 |

*Note that all times are in Universal Time (UT), otherwise known as Greenwich Mean Time (GMT). These times do not take Summer Time (BST) into account.*

# Moonrise and Moonset 2023

| Location | Date | Rise | Azimuth ° | Set | Azimuth ° |
|---|---|---|---|---|---|
| **Belfast** | | | | | |
| | Jul 01 (Sat) | 19:45 | 140 | 01:25 | 225 |
| | Jul 11 (Tue) | – | – | 14:56 | 296 |
| | Jul 21 (Fri) | 07:58 | 68 | 22:27 | 286 |
| | Jul 31 (Mon) | 20:35 | 140 | 01:40 | 217 |
| **Cardiff** | | | | | |
| | Jul 01 (Sat) | 19:09 | 136 | 01:34 | 229 |
| | Jul 11 (Tue) | – | – | 14:34 | 294 |
| | Jul 21 (Fri) | 07:55 | 70 | 22:10 | 285 |
| | Jul 31 (Mon) | 20:01 | 145 | 01:55 | 222 |
| **Edinburgh** | | | | | |
| | Jul 01 (Sat) | 19:46 | 143 | 01:04 | 223 |
| | Jul 11 (Tue) | 23:50 | 59 | 14:50 | 297 |
| | Jul 21 (Fri) | 07:43 | 67 | 22:19 | 287 |
| | Jul 31 (Mon) | 20:36 | 142 | 01:14 | 214 |
| **London** | | | | | |
| | Jul 01 (Sat) | 18:58 | 136 | 01:21 | 229 |
| | Jul 11 (Tue) | 23:55 | 62 | 14:22 | 294 |
| | Jul 21 (Fri) | 07:42 | 70 | 21:59 | 285 |
| | Jul 31 (Mon) | 19:50 | 135 | 01:42 | 222 |

*Note that all times are in Universal Time (UT), otherwise known as Greenwich Mean Time (GMT). These times do not take Summer Time (BST) into account.*

# Twilight Diagrams 2023

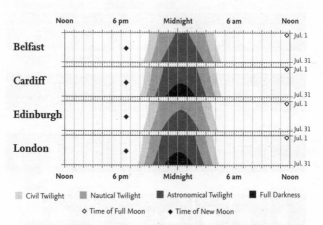

| | Civil Twilight | Nautical Twilight | Astronomical Twilight | Full Darkness |

◇ Time of Full Moon    ◆ Time of New Moon

**The exact times of the Moon's major phases are shown on the diagrams opposite.**

*Troposphere*
The lowest layer in the atmosphere, in which essentially all weather occurs. It is defined by the way in which temperature declines with height, and is bounded at the top by the level known as the tropopause.

# The Moon's Phases and Ages 2023

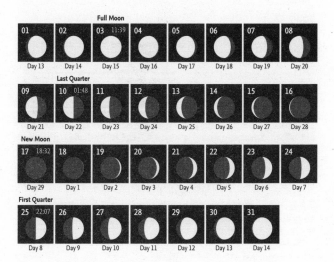

## Tropopause

The tropopause is the boundary between the lowest layer in the atmosphere (the troposphere) and the next highest (the stratosphere). It is an inversion at which temperatures either stabilise or begin to increase with height in the overlying stratosphere. The height of the tropopause (and thus the depth of the troposphere) increases from about 7 kilometres at the poles to 14–18 kilometres at the equator. There are breaks in the level of the tropopause (particularly near the location of jet streams) and these do allow some exchange of air between the layers.

# July – In this month

**3 July 1976** – During the 'Drought Summer' the temperature reached 35.9°C at Cheltenham in Gloucestershire on this day. The heatwave began in the first week of June and thunderstorms arrived on August 29.

**4 July 1862** – The heatwave today caused the Reverend Charles Dodgson (Lewis Carroll) to take his three nieces on the river Isis at Oxford. Stopping under a hayrick – the only shade – a request to 'tell us a story' resulted in the story of 'Alice in Wonderland'.

**5 July 1822** – For the first time ever, ordinary Londoners were able to buy ice cream. Previously, ice creams were available only to the well-to-do.

**7 July 1981** – The Toxteth Riots reached their peak on this day. Although the riots had been happening for four days under extremely hot conditions, they became even more violent. (The temperature reaches 31°C during the heatwave.) Although not exceptional, the temperature is thought to exacerbate the situation.

**8 July 1938** – A gale-force wind (31 m/s or 60 knots, 69 mph) hits the Royal St George's golf course, near Sandwich in Kent, on the final day of the British Open championship. Not only is a marquee lifted into the air, but one professional player takes 14 shots at the twelfth hole.

**9 July 1923 –** There was an extraordinary display of lightning over London. Some 6,924 strokes were recorded in six hours.

**9 July 1984 –** Lightning initiates a major fire at York Minster, which rages for hours, destroying much of Britain's largest medieval cathedral's roof, and the famous rose window. (Although much of the stained glass is salvageable.)

**11 July 1934 –** The temperature reached 33.3ºC in Nottinghamshire. It was the hottest day of a long hot spell. The summers of the two preceding years have also been hot and dry. The government passes the emergency Supply of Water in Bulk Act, allowing one water company to supply another.

**15 July 1998 –** A sudden gust or whirlwind – known locally as 'The Roger' or 'Sir Rodge's Blast' hits boats on the Norfolk Broads. It strikes without warning and is extremely localised; one vessel may be overturned and another nearby completely unaffected.

**17 July 1212 –** A strong southerly wind spread a fire through the timber houses in Southwark, London and an extensive area of the borough was razed to the ground. Some of the houses on London Bridge also burned. More than one thousand lives were lost.

# The Cloud

Percy Bysshe Shelley's poem *The Cloud* was also partly inspired by Howard's work and book (see pages 74 and 75). It was originally published in London in August 1820 as part of the collection that introduced Shelley's major work *Prometheus Unbound*. The poem *The Cloud* was one of the other poems that were included in the book.

**The Cloud**
*I bring fresh showers for the thirsting flowers,*
*From the seas and the streams;*
*I bear light shade for the leaves when laid*
*In their noonday dreams.*
*From my wings are shaken the dews that waken*
*The sweet buds every one,*
*When rocked to rest on their mother's breast,*
*As she dances about the sun.*
*I wield the flail of the lashing hail,*
*And whiten the green plains under,*
*And then again I dissolve it in rain,*
*And laugh as I pass in thunder.*

*I sift the snow on the mountains below,*
*And their great pines groan aghast;*
*And all the night 'tis my pillow white,*
*While I sleep in the arms of the blast.*
*Sublime on the towers of my skiey bowers,*
*Lightning my pilot sits;*
*In a cavern under is fettered the thunder,*
*It struggles and howls at fits;*
*Over earth and ocean, with gentle motion,*
*This pilot is guiding me,*
*Lured by the love of the genii that move*
*In the depths of the purple sea;*
*Over the rills, and the crags, and the hills,*
*Over the lakes and the plains,*
*Wherever he dream, under mountain or stream,*
*The Spirit he loves remains;*
*And I all the while bask in Heaven's blue smile,*
*Whilst he is dissolving in rains.*

The sanguine Sunrise, with his meteor eyes,
And his burning plumes outspread,
Leaps on the back of my sailing rack,
When the morning star shines dead;
As on the jag of a mountain crag,
Which an earthquake rocks and swings,
An eagle alit one moment may sit
In the light of its golden wings.
And when Sunset may breathe, from the lit sea beneath,
Its ardours of rest and of love,
And the crimson pall of eve may fall
From the depth of Heaven above,
With wings folded I rest, on mine aëry nest,
As still as a brooding dove.

That orbèd maiden with white fire laden,
Whom mortals call the Moon,
Glides glimmering o'er my fleece-like floor,
By the midnight breezes strewn;
And wherever the beat of her unseen feet,
Which only the angels hear,
May have broken the woof of my tent's thin roof,
The stars peep behind her and peer;
And I laugh to see them whirl and flee,
Like a swarm of golden bees,
When I widen the rent in my wind-built tent,
Till calm the rivers, lakes, and seas,
Like strips of the sky fallen through me on high,
Are each paved with the moon and these.

I bind the Sun's throne with a burning zone,
And the Moon's with a girdle of pearl;
The volcanoes are dim, and the stars reel and swim,
When the whirlwinds my banner unfurl.
From cape to cape, with a bridge-like shape,
Over a torrent sea,
Sunbeam-proof, I hang like a roof,
The mountains its columns be.
The triumphal arch through which I march

With hurricane, fire, and snow,
When the Powers of the air are chained to my chair,
Is the million-coloured bow;
The sphere-fire above its soft colours wove,
While the moist Earth was laughing below.

I am the daughter of Earth and Water,
And the nursling of the Sky;
I pass through the pores of the ocean and shores;
I change, but I cannot die.
For after the rain when with never a stain
The pavilion of Heaven is bare,
And the winds and sunbeams with their convex gleams
Build up the blue dome of air,
I silently laugh at my own cenotaph,
And out of the caverns of rain,
Like a child from the womb, like a ghost from the tomb,
I arise and unbuild it again.

August

# Introduction

Although August is always regarded as high summer, it sees the hottest day only as frequently as June. Both have fewer such days than July. August is also a quiet month, with less wind, being similar to July in that respect. However, to many counties in the east of England it is the wettest month as many thunderstorms and their associated rain travel across from the west.

The very first weather forecast appeared in *The Times* on 1 August 1861. It was prepared by Robert Fitzroy – best known to the general public as the captain of *HMS Beagle*, in which Charles Darwin made his momentous voyage. Fitzroy had been appointed in 1854 as 'Statist' to the newly formed Meteorological Department of the Board of Trade. Although, initially, Fitzroy merely collated observations submitted by mariners, he then instigated a system for communicating gale warnings to sailors at various ports. He went on to start 'prognosticating' forthcoming weather, and indeed introduced the term 'weather forecasting'. This resulted in his producing, and *The Times* publishing, the first ever forecast.

This revolutionary forecast was:

*General weather probable during next two days:*
*North – Moderate westerly wind; fine*
*West – Moderate south-westerly; fine*
*South – Fresh westerly; fine*

Unfortunately, British weather being so changeable and (in those days) so unpredictable, Fitzroy's forecasts soon became inaccurate and developed into a subject of derision. The situation was not helped by those meteorologists who wanted to put the subject on a scientific basis. Fitzroy's empirical methods were particularly criticised by Francis Galton who, in effect, wished to calculate everything. (Galton is infamous for proposing and promoting eugenics, and famous for producing an equation for obtaining the optimum cup of tea.) As the physics behind meteorology were so poorly understood at that time, Galton believed that true forecasts were impossible. Nevertheless, he remained fascinated by meteorology, and produced the first weather map, discovered anticyclones and proposed a theory explaining their existence.

Fitzroy's storm warning system was discontinued, but reinstated within a short period because both the general public and mariners campaigned for the reintroduction. The weather forecasts were also stopped, and only on 1 April 1875 did *The Times* publish the first weather map, prepared by Galton (see page 145), as distinct from the first weather forecast.

August was originally the sixth month (known as *sextilis*) of the ten-month Roman calendar (which omitted the winter months and began the year in March). It became the eighth month when, in about 700 BCE the months January and February were added at the beginning of the year. It originally had 29 days, but gained 2 days when Julius Caesar created the Julian calendar in 46 BCE. It was renamed in honour of the emperor Augustus in 8 BCE. It retained 31 days when Pope Gregory XIII instigated the calendar reform leading to the Gregorian calendar, introducing the new calendar in 1582.

Until 1965, the August Bank Holiday was held at the beginning (not the end) of the month. The first Monday in August is still a public holiday in Scotland, but in England, Wales and Northern Ireland it has been moved to the last Monday in August.

*Robert Fitzroy (1805–1865), widely regarded as the founder of modern meteorology. A sea area is named after him, and the headquarters of the Met Office are on Fitzroy Road in Exeter.*

# Weather Extremes

| Country | Temp. | Location | Date |
|---|---|---|---|
| *Maximum temperature* | | | |
| England | 38.5°C | Faversham (Kent) | 10 Aug. 2003 |
| Northern Ireland | 30.6°C | Tandragee Ballylisk (Co. Armagh) | 2 Aug. 1995 |
| Scotland | 32.9°C | Greycrook (Scottish Borders) | 9 Aug. 2005 |
| Wales | 35.2°C | Hawarden Bridge (Flintshire) | 2 Aug. 1990 |
| *Minimum temperature* | | | |
| England | -2.0°C | Moor House (Cumbria) Kielder Castle (Northumberland) | 28 Aug. 1977 14 Aug. 1994 |
| Northern Ireland | -1.9°C | Katesbridge (Co. Down) | 24 Aug. 2014 |
| Scotland | -4.5°C | Lagganlia (Inverness-shire) | 21 Aug. 1973 |
| Wales | -2.8°C | Alwen (Conwy) | 29 Aug. 1959 |

| Country | Pressure | Location | Date |
|---|---|---|---|
| *Maximum pressure* | | | |
| Scotland | 1037.4 hPa | Kirkwall (Orkney) | 25 Aug. 1968 |
| *Minimum pressure* | | | |
| Eire | 967.7 hPa | Belmullet (Co. Mayo) | 14 Aug. 1959 |

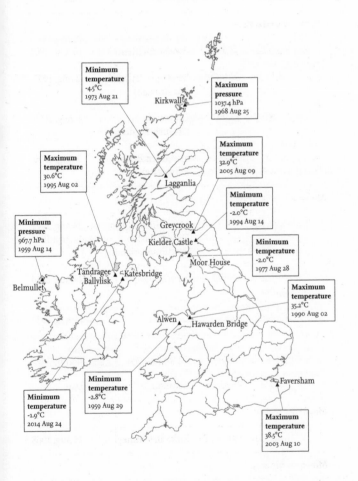

**Minimum temperature**
-4.5°C
1973 Aug 21

Kirkwall

**Maximum pressure**
1037.4 hPa
1968 Aug 25

**Maximum temperature**
30.6°C
1995 Aug 02

Lagganlia

**Maximum temperature**
32.9°C
2005 Aug 09

**Minimum temperature**
-2.0°C
1994 Aug 14

Greycrook

Kielder Castle

Moor House

**Minimum temperature**
-2.0°C
1977 Aug 28

**Minimum pressure**
967.7 hPa
1959 Aug 14

Belmullet

Tandragee
Ballylisk

Katesbridge

**Maximum temperature**
35.2°C
1990 Aug 02

Alwen
Hawarden Bridge

Faversham

**Minimum temperature**
-1.9°C
2014 Aug 24

**Minimum temperature**
-2.8°C
1959 Aug 29

**Maximum temperature**
38.5°C
2003 Aug 10

133

# The Weather in August 2021

| Observation | Location | Date |
| --- | --- | --- |
| *Max. temperature*<br>27.2°C | Tyndrum (Perthshire) | 25 August |
| *Min. temperature*<br>0.3°C | Braemar (Aberdeenshire) | 31 August |
| *Rainfall*<br>74.2 mm | Spadeadam (Cumbria) | 9 August |
| *Wind gust*<br>26 m/s<br>(51 knots or 59 mph) | Needles Old Battery<br>(Isle of Wight) | 9 August |

*Ionosphere*

A region of the atmosphere, consisting of the upper mesosphere and part of the exosphere (from about 60–70 km to 1000 km or more) where radiation from the Sun ionises atoms and causes high electrical conductivity. The ionosphere both reflects certain radio waves back towards the surface, and blocks some wavelengths of radiation from space.

The extreme weather at the end of July 2021 continued into August, particularly in the south and south-east. There was flooding in Kent on August 1, and disruption by floods to both rail and road traffic in South Wales (and also on the Isle of Wight) on August 2. Two days later, on August 4, there was rail disruption in Northern Ireland and also across a band of northern England. High winds were experienced in western Scotland and in the Northern Isles. Heavy rain affected Wales in successive days, with Capel Curig in Gwyedd recording 53.0 mm on August 5. Heavy bands of rain continued to affect the country until August 9. The wettest areas were the south-east of England and southern Scotland, where there was flooding in both the Edinburgh and Glasgow areas on August 6, with road and rail disruption. The heavy rain in Scotland then spread north and east. There were floods in Northern Ireland and severe winds and flooding in East Anglia. There were widespread incidents of flooding in southern England, from August 7, with thunderstorms affecting all southern and eastern counties. Parts of the south became very windy, with a maximum gust of 26.2 m/s (51 knots or 59 mph) recorded at Needles Old Battery on the Isle of Wight on August 9. On the same date, Spadeadam in Cumbria recorded the month's rainfall maximum of 74.2 mm.

Sunshine was low, and for England, the month proved to be the seventh dullest August in a series from 1919. East Anglia became warmer from mid-month, with Cavendish in Suffolk reaching a temperature of 26.4°C on August 15.

Towards the end of the month, fine conditions returned to Scotland, with the month's maximum temperature of 27.2°C recorded at Tyndrum in Perthshire on August 25. The monthly minimum temperature of 0.3°C was experienced at Braemar in Aberdeenshire on the night of August 31.

# Sunrise and Sunset 2023

| Location | Date | Rise | Azimuth ° | Set | Azimuth ° |
|----------|------|------|-----------|-----|-----------|
| **Belfast** | | | | | |
| | Aug 01 (Tue) | 04:34 | 56 | 20:25 | 303 |
| | Aug 11 (Fri) | 04:52 | 61 | 20:05 | 298 |
| | Aug 21 (Mon) | 05:10 | 67 | 19:43 | 292 |
| | Aug 31 (Thu) | 05:28 | 74 | 19:19 | 286 |
| **Cardiff** | | | | | |
| | Aug 01 (Tue) | 04:36 | 59 | 20:01 | 301 |
| | Aug 11 (Fri) | 04:51 | 64 | 19:43 | 296 |
| | Aug 21 (Mon) | 05:07 | 69 | 19:23 | 291 |
| | Aug 31 (Thu) | 05:23 | 75 | 19:02 | 285 |
| **Edinburgh** | | | | | |
| | Aug 01 (Tue) | 04:17 | 55 | 20:20 | 305 |
| | Aug 11 (Fri) | 04:36 | 60 | 19:59 | 299 |
| | Aug 21 (Mon) | 04:55 | 67 | 19:36 | 293 |
| | Aug 31 (Thu) | 05:15 | 73 | 19:10 | 287 |
| **London** | | | | | |
| | Aug 01 (Tue) | 04:24 | 59 | 19:50 | 301 |
| | Aug 11 (Fri) | 04:39 | 64 | 19:32 | 296 |
| | Aug 21 (Mon) | 04:55 | 69 | 19:12 | 291 |
| | Aug 31 (Thu) | 05:112 | 75 | 18:50 | 285 |

*Note that all times are in Universal Time (UT), otherwise known as Greenwich Mean Time (GMT). These times do not take Summer Time (BST) into account.*

# Moonrise and Moonset 2023

| Location | Date | Rise | Azimuth ° | Set | Azimuth ° |
|----------|------|------|-----------|-----|-----------|
| **Belfast** | | | | | |
| | Aug 01 (Tue) | 21:04 | 131 | 03:10 | 222 |
| | Aug 11 (Fri) | 23:54 | 36 | 18:03 | 323 |
| | Aug 21 (Mon) | 10:45 | 106 | 21:05 | 250 |
| | Aug 31 (Thu) | 19:51 | 103 | 05:32 | 250 |
| **Cardiff** | | | | | |
| | Aug 01 (Tue) | 20:35 | 128 | 03:21 | 227 |
| | Aug 11 (Fri) | – | – | 17:25 | 318 |
| | Aug 21 (Mon) | 10:28 | 104 | 21:01 | 252 |
| | Aug 31 (Thu) | 19:36 | 102 | 05:28 | 251 |
| **Edinburgh** | | | | | |
| | Aug 01 (Tue) | 21:01 | 133 | 02:47 | 220 |
| | Aug 11 (Fri) | 23:28 | 34 | 18:05 | 326 |
| | Aug 21 (Mon) | 10:36 | 106 | 20:51 | 250 |
| | Aug 31 (Thu) | 19:42 | 103 | 05:17 | 249 |
| **London** | | | | | |
| | Aug 01 (Tue) | 20:24 | 128 | 03:07 | 226 |
| | Aug 11 (Fri) | 23:56 | 41 | 17:14 | 318 |
| | Aug 21 (Mon) | 10:15 | 104 | 20:49 | 252 |
| | Aug 31 (Thu) | 19:24 | 102 | 05:15 | 251 |

*Note that all times are in Universal Time (UT), otherwise known as Greenwich Mean Time (GMT). These times do not take Summer Time (BST) into account.*

# Twilight Diagrams 2023

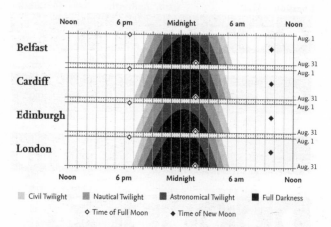

| | Civil Twilight | | Nautical Twilight | | Astronomical Twilight | | Full Darkness |
| --- | --- | --- | --- | --- | --- | --- | --- |

◇ Time of Full Moon     ◆ Time of New Moon

**The exact times of the Moon's major phases are shown on the diagrams opposite.**

......................................................................................................

### Föhn effect

When humid air is forced to rise over high ground, it normally deposits some precipitation in the form of rain or snow. When the air descends on the far side of the hills, because it has lost some of its moisture, it warms at a greater rate than it cooled on its ascent. This 'föhn effect' may cause temperatures on the leeward side of hills or mountains to be much warmer than locations at a corresponding altitude on the windward side.

......................................................................................................

# The Moon's Phases and Ages 2023

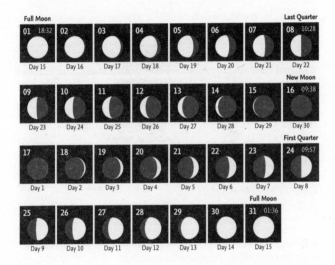

**Full Moon**

| 01 18:32 | 02 | 03 | 04 | 05 | 06 | 07 | 08 10:28 |
|---|---|---|---|---|---|---|---|
| Day 15 | Day 16 | Day 17 | Day 18 | Day 19 | Day 20 | Day 21 | Day 22 |

**Last Quarter**

**New Moon**

| 09 | 10 | 11 | 12 | 13 | 14 | 15 | 16 09:38 |
|---|---|---|---|---|---|---|---|
| Day 23 | Day 24 | Day 25 | Day 26 | Day 27 | Day 28 | Day 29 | Day 30 |

**First Quarter**

| 17 | 18 | 19 | 20 | 21 | 22 | 23 | 24 09:57 |
|---|---|---|---|---|---|---|---|
| Day 1 | Day 2 | Day 3 | Day 4 | Day 5 | Day 6 | Day 7 | Day 8 |

**Full Moon**

| 25 | 26 | 27 | 28 | 29 | 30 | 31 01:36 |
|---|---|---|---|---|---|---|
| Day 9 | Day 10 | Day 11 | Day 12 | Day 13 | Day 14 | Day 15 |

---

*Advection*
The horizontal motion of air from one area to another. Advection of humid air over cold ground will often result in the ground cooling the overlying air, and is a common cause of mist and fog. Advection also brings sea mist onto low-lying coastal areas.

---

# August – In this month

**1 August 1861 –** The very first public weather forecast, prepared by Robert Fitzroy, was published in *The Times* newspaper (see page 144).

**1 August 1846 –** A day of hail. An extreme hailstorm over London shatters over 7,000 panes of glass in the Houses of Parliament, 10,000 in Leicester Square, and almost every pane in the Burlington Arcade, Regent Street Arcade and at Scotland Yard, Somerset House and the picture gallery at Buckingham Palace.

**3 August 1879 –** Another day of hail. Some 3,000 panes are broken at the Temperate House at Kew Gardens and 700 in the Great Palm House.

**4 August 1666 –** The wind prevented the Dutch fleet from affecting a landing at the Medway, to fire on the English fleet in port. They blockaded the Thames instead.

**5 August 2003 –** Network Rail impose a speed restriction of 60 mph (96 kph) on all railway lines in the Midlands and southern England as temperatures reach slightly more than 30°C. Newspaper headlines proclaim 'Britain's Railways blame the wrong kind of sunshine'.

**6 August 1956 –** On August Bank Holiday, some 1.2 metres of hail fall on Tunbridge Wells in Kent.

**6 August 1981 –** The sky over London became black. Street lights came on automatically, and motorists had to use headlights. Torrential rain followed. A giant cumulonimbus cloud had formed, thought to be at least 13 km high.

**7 August 1829** – A Hebridean storm during a trip to the island of Staffa, inspires the young Felix Mendelssohn to write the opening theme to his famous concert overture 'The Hebrides'. This was not published until 1833.

**9 August 1843** – An extreme hailstorm, accompanied by lightning and torrential rain crossed the country leaving a 'hail swathe' from Horsey in Norfolk to near Stow-on-the-Wold in Gloucestershire, probably accompanying a tornado. It was almost certainly caused by a rare supercell storm and is one of the most destructive recorded in Britain. It is ranked H7 on the TORRO hail scale. There was major destruction all along the storm's track.

**14 August 1979** – An extremely long-lasting rainbow was observed from the coast of North Wales. This lasted for three hours rather than the usual few minutes. Some have tried to find a connection with the disastrous storm that occurred during the Fastnet Race that day.

**20 August 1860** – The Church of England issued a 'Prayer for Fair Weather' in response to persistent rain throughout the summer of 1860, which ruined crops. Luckily, the weather improves and the autumn had below average rainfall, so part of the harvest was salvaged.

**26 August 1912** – The Great Norfolk Flood resulted from torrential rain after a month of heavy rain that caused the ground to be waterlogged. Three times the normal amount of monthly rain occurred in a single day and some 40 bridges were washed away. Norwich was completely cut off by road and rail.

**29 August 1561** – A haar, the sea-fog of Britain's North Sea coasts, saved Mary, Queen of Scots from capture at the end of her trip from France. She lands undetected at Leith.

# Drought

Droughts are relatively common in Britain. With a high population density and extensive agriculture, a water shortage is often experienced. Although the definition of a drought varies from country to country, meteorologically a drought is considered to apply when three weeks occur with less than one third of normal rainfall for the time of year.

In Britain, since 1990, a drought is officially defined as 'a period of at least 15 consecutive days when there is less than 0.2 mm (0.0008 inches) of rainfall'. In the longer term, drought may also be defined as 'a 50% deficit over three months, or a 15% shortfall over two years'. The south-east of England tends to suffer most in any drought, because of its high population density (and, thus, high domestic demand), extensive, intensive agriculture, and generally low precipitation.

A drought normally occurs when a 'blocking high' forms, diverting rain-bearing depressions, generally southwards. Usually, the Mediterranean area receives additional rain, and may experience a 'bad' summer.

The definitions just given tend to apply to the demand for domestic water supplies. A different form of drought is a hydrological or agricultural drought. Here, moisture in the soil is not reaching plants, either because it is frozen (which does occur in Britain in severe winters) or because high temperatures raise the rate of evapotranspiration (loss of water from plants by evaporation). Such droughts may occur after a dry winter, because water takes time to percolate downwards and recharge underground reservoirs and the water table. Such conditions may persist for several years, even with high amounts of rainfall, simply because of the time required for recharge. An agricultural drought is sometimes defined as a shortfall of rain during the growing season.

One particularly notable British drought was that of 1934, when an emergency bill was rushed through Parliament: the 'Supply of Water in Bulk Act'. That allowed water companies, previously restricted to a single area, to supply other water boards with water.

The summer of 1976 is sometimes known as the 'Drought Summer'. It was one of the most severe droughts on record.

A very hot summer followed a dry winter 1975–76. Although the drought effectively started in October 1975, it did not present a serious problem until late spring 1976. In April no rain fell in parts of the West Country or in Kent. In Devon and Dorset some locations recorded no rain for 45 consecutive days in July and August. This was in addition to three periods of absolute drought with no measurable precipitation for 58 days. The conditions were so severe that the government brought in the fourth emergency stage of their drought policy, with drastic water rationing, with no supplies whatsoever to domestic consumers (usually during specific times of day), and the use of standpipes and water tankers.

In late August 1976, the politician Denis Howell was appointed 'Minister for Drought' (rapidly nicknamed 'Minister for Rain'). Heavy rains a few days later brought extensive flooding, with heavy rainfall onto parched, hardened soil. His title was then changed to 'Minister of Floods'. Later, during the severe winter of 1978–79, the same person was appointed 'Minister for Snow'.

*The Upper Neuadd Reservoir in the intense drought of 1976.*

Below: *The very first weather forecast, prepared by Robert Fitzroy, and published in* The Times *of 1 August 1861.*

## THE WEATHER.

### METEOROLOGICAL REPORTS.

| Wednesday, July 31, 8 to 9 a.m. | B. | E. | M. | D. | F. | C. | I. | S. |
|---|---|---|---|---|---|---|---|---|
| Nairn.. .. .. | 29·54 | 57 | 56 | W.S.W. | 6 | 9 | o. | 3 |
| Aberdeen .. .. | 29·60 | 59 | 54 | S.S.W. | 5 | 1 | b. | 3 |
| Leith .. .. .. | 29·70 | 61 | 55 | W. | 3 | 5 | c. | 2 |
| Berwick .. .. | 29·69 | 59 | 55 | W.S.W. | 4 | 4 | c. | 2 |
| Ardrossan .. | 29·73 | 57 | 55 | W. | 5 | 4 | c. | 5 |
| Portrush .. .. | 29·72 | 57 | 54 | S.W. | 2 | 2 | b. | 2 |
| Shields .. .. | 29·80 | 59 | 54 | W.S.W. | 4 | 5 | o. | 3 |
| Galway .. .. | 29·63 | 65 | 62 | W. | 5 | 4 | c. | 4 |
| Scarborough .. | 29·85 | 59 | 56 | W. | 3 | 6 | c. | 2 |
| Liverpool .. .. | 29·91 | 61 | 56 | S.W. | 2 | 8 | c. | 2 |
| Valentia .. .. | 29·37 | 62 | 60 | S.W. | 2 | 5 | o. | 3 |
| Queenstown .. | 29·88 | 61 | 59 | W. | 3 | 5 | c. | 2 |
| Yarmouth.. .. | 30·05 | 61 | 59 | W. | 5 | 2 | c. | 3 |
| London .. .. | 30·02 | 62 | 56 | S.W. | 3 | 2 | b. | — |
| Dover.. .. .. | 30·04 | 70 | 61 | S.W. | 3 | 7 | o. | 2 |
| Portsmouth .. | 30·01 | 61 | 59 | W. | 3 | 6 | o. | 2 |
| Portland .. .. | 30·03 | 63 | 59 | S.W. | 3 | 2 | c. | 3 |
| Plymouth.. .. | 30·00 | 62 | 59 | W. | 5 | 1 | b. | 4 |
| Penzance .. .. | 30·04 | 61 | 60 | S.W. | 2 | 6 | c. | 3 |
| Copenhagen .. | 29·94 | 64 | — | W.S.W. | 2 | 6 | c. | 3 |
| Helder .. .. | 29·99 | 63 | — | W.S.W. | 6 | 5 | c. | 3 |
| Brest .. .. | 30·09 | 60 | — | S.W. | 2 | 6 | c. | 5 |
| Bayonne .. .. | 30·13 | 68 | — | — | — | 9 | m. | 5 |
| Lisbon .. .. | 30·18 | 70 | — | N.N.W. | 4 | 3 | b. | 2 |

*General* weather probable during next two days in the—

North—Moderate westerly wind ; fine.

West—Moderate south-westerly ; fine.

South—Fresh westerly ; fine.

Explanation.

B. Barometer, corrected and reduced to 32° at mean sea level ; each 10 feet of vertical rise causing about one-hundredth of an inch diminution, and each 10° above 32° causing nearly three-hundredths increase. E. Exposed thermometer in shade. M. Moistened bulb (for evaporation and dew-point). D. Direction of wind (true—two points *left* of magnetic). F. Force (1 to 12—estimated). C. Cloud (1 to 9). I. Initials :—b., blue sky ; c., clouds (detached) ; f., fog ; h., hail ; l, lightning ; m., misty (hazy) ; o., overcast (dull) ; r., rain ; s., snow ; t., thunder. S. Sea disturbance (1 to 9).

Next page: *The first weather map, prepared by Francis Galton, and published in* The Times *of 1 April 1875, 14 years after Fitzroy's forecast.*

WEATHER CHART, MARCH 31, 1875.

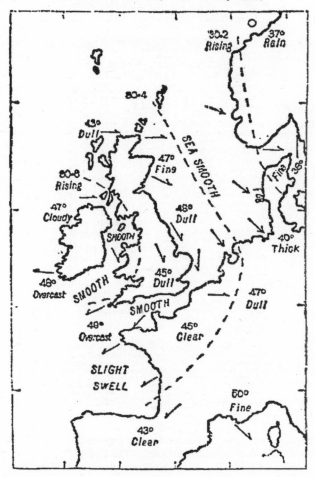

September

# Introduction

September has always been regarded as a transitional month, between the high summer of July and August and the autumn. It is a time for late holidays and generally fairly quiet weather. The month tends to be drier than the preceding month, August, and the following one, October. Although there is often a period of windy weather, sometimes with gale-force winds, generally late in the month, the mariners' myth of 'equinoctial gales' certainly does not apply (see the discussion of this point under March, page 46).

To meteorologists, September is the first month of autumn. However, if the year is regarded as consisting of five seasons, the transition to autumn occurs a bit later, after the first week of the month, in which the weather has often been as warm as in June. Because the month includes the equinox (September 23 in 2023), the length of daylight begins to shorten noticeably, and the longer time after sunset allows evening and overnight temperatures to fall. In quiet weather, temperatures soon fall to the saturation point, and dew forms on the ground and vegetation. Temperatures fall even farther overnight, so that

---

*Autumn – September 10 to November 19*
This season is marked by an early period when settled weather may occur, as high pressure makes an occasional incursion across the country from the south or south-west. It is, however, usually marked by a period of wet and windy weather, although this tends to die down as the season transitions in the middle of November into that of the early winter.

---

there is often some morning mist and fog. It was in mid-September that John Keats composed his well-loved poem 'To Autumn', with its famous line of 'Season of mists and mellow fruitfulness'.

During the summer, the air arriving from the Atlantic is of the type known to meteorologists as tropical maritime air: cool and fairly humid. The depressions arriving over the country from the Atlantic are fairly shallow, and their winds are subdued. In September, colder air (polar maritime air) begins to come south, and the greater temperature contrast relative to the warmer air from the south tends to deepen depressions and increase their wind speeds.

In Europe, the Full Moon in September is generally known as the 'Harvest Moon'. This is traditionally the Full Moon closest to the equinox. It is also because this is normally the time of the greatest harvest, not only of the various cereal grains, such as wheat and barley, but also apples and similar tree fruits. European harvest festivals were generally held on the Sunday closest to the Full Moon in September.

### Stratosphere

The second layer in the atmosphere, lying above the troposphere, in which temperatures either stabilise or begin to increase with height. This increase of temperature is primarily driven by the absorption of solar energy by ozone in the ozone layer. In the lowermost region, between the tropopause and about a height of 20 kilometres, the temperature is stable. Above that there is an overall increase to the top of the stratosphere (the stratopause) at an altitude of about 50 kilometres.

# Weather Extremes

| Country | Temp. | Location | Date |
|---|---|---|---|
| *Maximum temperature* | | | |
| England | 35.6°C | Bawtry – Hesley Hall (South Yorkshire) | 2 Sep. 1906 |
| Northern Ireland | 27.8°C | Armagh (Co. Armagh) | 1 Sep. 1906 |
| Scotland | 32.2°C | Gordon Castle (Moray) | 1 Sep. 1906 |
| Wales | 31.1°C | Gogerddan (Powys) | 1 Sep. 1961 |
| *Minimum temperature* | | | |
| England | -5.6°C | Stanton Downham (Norfolk) Grendon Underwood (Buckinghamshire) | 30 Sep. 1969 |
| Northern Ireland | -3.2°C | Magherally (Co. Down) | 30 Sep. 1991 |
| Scotland | -6.7°C | Dalwhinnie (Inverness-shire) | 26 Sep. 1942 |
| Wales | -5.5°C | St Harmon (Powys) | 19 Sep. 1986 |

| Country | Pressure | Location | Date |
|---|---|---|---|
| *Maximum pressure* | | | |
| Northern Ireland | 1043 hPa | Ballykelly (Co. Londonderry) | 11 Sep. 2009 |
| *Minimum pressure* | | | |
| Eire | 957.1 hPa | Claremorris (Co. Mayo) | 21 Sep. 1953 |

Maximum
temperature
32.2°C
1906 Sep 01

Gordon Castle

Minimum
temperature
-6.7°C
1942 Sep 26

Maximum
pressure
1043 hPa
2009 Sep 11

Dalwhinnie

Minimum
temperature
-3.2°C
1991 Sep 30

Maximum
temperature
27.8°C
1906 Sep 01

Ballykelly

Maximum
temperature
35.6°C
1906 Sep 02

Armagh   Magherally

Claremorris

Bawtry – Hesley Hall

Stanton Downham

Gogerddan   St Harmon

Minimum
pressure
957.1 hPa
1953 Sep 21

Grendon
Underwood

Maximum
temperature
31.1°C
1961 Sep 01

Minimum
temperature
-5.6°C
1969 Sep 30

Minimum
temperature
-5.5°C
1986 Sep 19

# The Weather in September 2021

| Observation | Location | Date |
| --- | --- | --- |
| **Max. temperature** 30.8°C | Hartpury College (Gloucestershire) | 7 September |
| **Min. temperature** -1.4°C | Aboyne (Aberdeenshire) | 30 September |
| **Rainfall** 76.6 mm | White Barrow (Devon) | 29 September |
| **Wind gust** 33 m/s (64 knots or 74 mph) | Needles Old Battery (Isle of Wight) | 17 September |

Although much of the weather in September 2021 was settled, this was broken in the south-east of England on September 8, when there were thunderstorms in Kent. Two days later, lightning strikes caused disruption to railway signalling and some localised damage. There were also various instances of flooding in the south-east. In north Wales, in Flintshire, there was flooding from torrential rain, as well as floods over the border in England, in northern Cheshire.

In the middle of the month (September 14) there was severe weather in the eastern counties of England. There were also roads flooded in London, delays on several of the Underground lines and even closure of the Blackwall Tunnel because of flooding. More flooding occurred on September 19 in Melford and Sudbury in Suffolk.

Strong winds affected the south of England and a gust of 33 m/s (64 knots or 74 mph) was recorded at Needles Old Battery on September 17.

The last week of the month saw very heavy rain arrive from the west, which affected most of England. White Barrow in Devon recorded 76.6 mm of rain in the 24 hours to 9:00 am on September 29. Kielder Castle in Northumberland recorded a low of 0.7°C on September 30 and showers and longer periods of rain affected all parts of England.

Northern Ireland began the month quietly, then increasing amounts of rain invaded the country. A reasonably quiet spell followed, but the weather turned wet and windy at the end of the month, with showers and thunderstorms. The last day of the month (September 30) saw heavy rain bands that affected all parts of the country.

Wales was wet and windy during the end of the month, and Scotland saw very heavy rain, with a fall of 37.4 mm at Tyndrum in Aberdeenshire on September 26. There were heavy showers spreading east on September 29. The weather was cold and Aboyne in Aberdeenshire recorded the month's low temperature of -1.4°C on September 30.

# Sunrise and Sunset 2023

| Location | Date | Rise | Azimuth ° | Set | Azimuth ° |
|---|---|---|---|---|---|
| **Belfast** | | | | | |
| | Sep 01 (Fri) | 05:30 | 76 | 19:16 | 285 |
| | Sep 11 (Mon) | 05:49 | 81 | 18:51 | 279 |
| | Sep 21 (Thu) | 06:07 | 87 | 18:26 | 272 |
| | Sep 30 (Sat) | 06:23 | 94 | 18:03 | 265 |
| **Cardiff** | | | | | |
| | Sep 01 (Fri) | 05:25 | 75 | 19:00 | 284 |
| | Sep 11 (Mon) | 05:41 | 81 | 18:37 | 278 |
| | Sep 21 (Thu) | 05:57 | 88 | 18:14 | 272 |
| | Sep 30 (Sat) | 06:11 | 93 | 17:53 | 266 |
| **Edinburgh** | | | | | |
| | Sep 01 (Fri) | 05:17 | 74 | 19:08 | 286 |
| | Sep 11 (Mon) | 05:36 | 80 | 18:42 | 279 |
| | Sep 21 (Thu) | 05:55 | 87 | 18:15 | 272 |
| | Sep 30 (Sat) | 06:13 | 94 | 17:52 | 266 |
| **London** | | | | | |
| | Sep 01 (Fri) | 05:13 | 75 | 18:48 | 284 |
| | Sep 11 (Mon) | 05:29 | 81 | 18:26 | 278 |
| | Sep 21 (Thu) | 05:45 | 88 | 18:02 | 272 |
| | Sep 30 (Sat) | 05:59 | 93 | 17:42 | 266 |

*Note that all times are in Universal Time (UT), otherwise known as Greenwich Mean Time (GMT). These times do not take Summer Time (BST) into account.*

# Moonrise and Moonset 2023

| Location | Date | Rise | Azimuth ° | Set | Azimuth ° |
|---|---|---|---|---|---|
| **Belfast** | | | | | |
| | Sep 01 (Fri) | 20:01 | 90 | 07:10 | 263 |
| | Sep 11 (Mon) | 01:00 | 45 | 18:24 | 310 |
| | Sep 21 (Thu) | 14:04 | 140 | 20:18 | 219 |
| | Sep 30 (Sat) | 18:26 | 73 | 07:40 | 282 |
| **Cardiff** | | | | | |
| | Sep 01 (Fri) | 19:49 | 90 | 07:00 | 263 |
| | Sep 11 (Mon) | 01:09 | 49 | 17:57 | 306 |
| | Sep 21 (Thu) | 13:29 | 135 | 20:32 | 223 |
| | Sep 30 (Sat) | 18:21 | 74 | 07:24 | 281 |
| **Edinburgh** | | | | | |
| | Sep 01 (Fri) | 19:50 | 91 | 06:57 | 263 |
| | Sep 11 (Mon) | 00:38 | 43 | 18:21 | 311 |
| | Sep 21 (Thu) | 14:05 | 142 | 19:55 | 216 |
| | Sep 30 (Sat) | 18:12 | 72 | 07:30 | 282 |
| **London** | | | | | |
| | Sep 01 (Fri) | 19:38 | 91 | 06:48 | 263 |
| | Sep 11 (Mon) | 00:56 | 49 | 17:45 | 307 |
| | Sep 21 (Thu) | 13:17 | 135 | 20:19 | 223 |
| | Sep 30 (Sat) | 18:09 | 74 | 07:11 | 281 |

*Note that all times are in Universal Time (UT), otherwise known as Greenwich Mean Time (GMT). These times do not take Summer Time (BST) into account.*

# Twilight Diagrams 2023

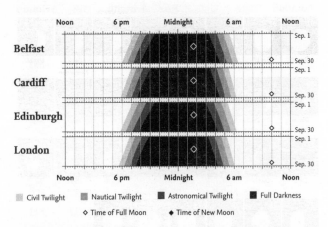

|  | Noon | 6 pm | Midnight | 6 am | Noon |
|--|------|------|----------|------|------|

Belfast

Cardiff

Edinburgh

London

Sep. 1 / Sep. 30 (for each city)

▨ Civil Twilight   ▨ Nautical Twilight   ■ Astronomical Twilight   ■ Full Darkness

◇ Time of Full Moon   ◆ Time of New Moon

**The exact times of the Moon's major phases are shown on the diagrams opposite.**

---

*Mesosphere*

The third layer of the atmosphere, above the stratosphere and below the thermosphere. It extends from about 50 km (the height of the stratopause) to about 86–100 km (the mesopause). Within it, temperature decreases with increasing altitude, reaching the atmospheric minimum of approximately -123°C at the mesopause. The only clouds occurring within the mesosphere are noctilucent clouds (see pages 108 and 109).

# The Moon's Phases and Ages 2023

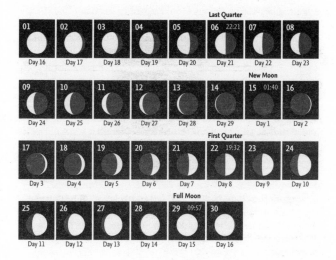

| | | | | | Last Quarter | | |
|---|---|---|---|---|---|---|---|
| 01 | 02 | 03 | 04 | 05 | 06 22:21 | 07 | 08 |
| Day 16 | Day 17 | Day 18 | Day 19 | Day 20 | Day 21 | Day 22 | Day 23 |
| | | | | | | New Moon | |
| 09 | 10 | 11 | 12 | 13 | 14 | 15 01:40 | 16 |
| Day 24 | Day 25 | Day 26 | Day 27 | Day 28 | Day 29 | Day 1 | Day 2 |
| | | | | | First Quarter | | |
| 17 | 18 | 19 | 20 | 21 | 22 19:32 | 23 | 24 |
| Day 3 | Day 4 | Day 5 | Day 6 | Day 7 | Day 8 | Day 9 | Day 10 |
| | | | Full Moon | | | | |
| 25 | 26 | 27 | 28 | 29 09:57 | 30 | | |
| Day 11 | Day 12 | Day 13 | Day 14 | Day 15 | Day 16 | | |

*Thermosphere*
The fourth layer of the atmosphere, counting from the surface.
It is tenuous and lies above the upper limit of the mesosphere,
the mesopause, at approximately 86–100 km, and extends
into interplanetary space. Within it, the temperature increases
continuously with height.

# September – In this month

**2 September 1667** – Robert Hooke noted in his diary that it rained on this day. He had previously noted that it rained heavily on St Swithin's Day, July 15, but that 40 days of drought followed, until it rained again on September 2.

**2 September 1906** – A temperature of 35.6°C was recorded at Bawtry in Yorkshire. This was the hottest temperature for 1906 and stands as an unbroken record for September.

**3 September 1928** – Alexander Fleming, the bacteriologist, tidying up the chaos that he had left in his laboratory when he went on holiday, noticed a mould that had blown in through the open window. The mould has killed some of the bacteria in the culture dish. *Penicillin*, the first antibiotic, had been discovered.

**4 September 1966** – Australians Jack Lydgate, Johnny McIlroy and Peter Russell become the first surfers to conquer The Cribbar, a huge wave (up to nearly 7 metres high) off Newquay in Cornwall, caused by the swell from Hurricane Faith.

**8 September 1845** – The *Dublin Evening Post* revealed that potato blight had arrived in Ireland. Potatoes thrived in the Irish soil and weather, essentially producing an agriculture that relied on a single crop. The weather proves ideal for the propagation of the fungus responsible and the Great Irish Potato Famine begins. It lasts until 1849.

**12 September 1666** – The Great Fire of London began in Pudding Lane. The wooden houses are tinder-dry after a drought lasting several weeks. The fire burns until September 14, fanned by an east wind. Remarkably, there were no known casualties. The wind signalled the fire's advance, so residents were able to escape.

**18 September 1993** – As much as 243.5 mm of rain fell at Cloore Lake in County Kerry in southern Ireland. This is a daily record for the whole of Ireland.

**19 September 1819** – A walk in the water meadows in Winchester inspired John Keats to write his ode 'To Autumn' (see pages 160 and 161).

**26 September 1950** – A blue-coloured sun, followed by a blue-coloured Moon were observed from Melrose in Scotland. Smoke from vast forest fires in Canada was responsible.

**29 September 1714** – A dense fog prevented George, Elector of Hanover, from landing at Greenwich at the intended time. Unfortunately, the courtiers awaiting his arrival to take up the crown of England had grown tired of waiting and dispersed. He disembarked at 6.00 pm to a deserted quayside.

# To Autumn

John Keats wrote his ode 'To Autumn' in September 1819. It was inspired by a walk that Keats took in the water meadows along the River Itchen at Winchester on 19 September 1819. It was actually the very last poem that he wrote, because other matters, including his declining health, then took up his time and creative energy until his death from tuberculosis in Rome about a year later.

The poem was published in 1820 in a collection entitled *Lamia, Isabella, the Eve of St. Agnes, and Other Poems.* It is one of the most famous and quoted odes (especially the first line) in the English language.

*A posthumous portrait of John Keats, painted by William Hilton, (c. 1822), now in the National Portrait Gallery, London.*

## To Autumn

*Season of mists and mellow fruitfulness,*
*Close bosom-friend of the maturing sun;*
*Conspiring with him how to load and bless*
*With fruit the vines that round the thatch-eves run;*
*To bend with apples the moss'd cottage-trees,*
*And fill all fruit with ripeness to the core;*
*To swell the gourd, and plump the hazel shells*
*With a sweet kernel; to set budding more,*
*And still more, later flowers for the bees,*
*Until they think warm days will never cease,*
*For summer has o'er-brimm'd their clammy cells.*

*Who hath not seen thee oft amid thy store?*
*Sometimes whoever seeks abroad may find*
*Thee sitting careless on a granary floor,*
*Thy hair soft-lifted by the winnowing wind;*
*Or on a half-reap'd furrow sound asleep,*
*Drows'd with the fume of poppies, while thy hook*
*Spares the next swath and all its twined flowers:*
*And sometimes like a gleaner thou dost keep*
*Steady thy laden head across a brook;*
*Or by a cyder-press, with patient look,*
*Thou watchest the last oozings hours by hours.*

*Where are the songs of spring? Ay, Where are they?*
*Think not of them, thou hast thy music too,–*
*While barred clouds bloom the soft-dying day,*
*And touch the stubble-plains with rosy hue;*
*Then in a wailful choir the small gnats mourn*
*Among the river sallows, borne aloft*
*Or sinking as the light wind lives or dies;*
*And full-grown lambs loud bleat from hilly bourn;*
*Hedge-crickets sing; and now with treble soft*
*The red-breast whistles from a garden-croft;*
*And gathering swallows twitter in the skies.*

October

# Introduction

October is definitely an autumnal month. It generally sees an increase in the number of depressions advancing across the country from the Atlantic. These are often vigorous, with strong winds and carrying plenty of rain. The nature of the accompanying weather depends on the type of air within the depressions. Frequently this will include warm, moist maritime tropical air, and give rise to dull days, often accompanied by extensive rain, because the air has passed over a relatively warm sea and thus taken up significant amounts of moisture. The unstable air behind the cold front of depressions often forms showers.

Because the Arctic is now rapidly cooling, any air arriving from the north may be extremely cold. When there is an incursion of frigid maritime Arctic air, the weather may include the first snow, as well as producing significant heavy showers, which may turn thundery and even turn into hailstorms. Scotland tends to experience more thunderstorms at this time, when strong showers arise in the unstable maritime Arctic air behind the cold fronts of depressions, but the air in front of these depressions arises from a warm sea. It may be heavily laden with moisture and give very heavy and prolonged rain ahead of the warm fronts.

Occasionally during the month there may be a quiet anticyclonic period, giving weather that resembles that in September. Generally the month sees the first extensive frosts, except, perhaps, in the very south of England. In Scotland, the trees may lose their leaves in the strong winds without a major display of colour. In England, the winds are often weaker, being farther from the centres of strong depressions, and so the leaves tend to persist on the trees and may provide a brilliant show of colour.

In Europe and Britain, the Full Moon in October was frequently called the 'Hunter's Moon'. This was the time when people prepared for the coming winter, both by hunting game and by slaughtering livestock. Every three years, however, the first Full Moon after the autumnal equinox actually fell in October. It was then customary to call that particular Full Moon the 'Harvest (rather than Hunter's) Moon'. Among the various tribes in North America, there was a tendency to name the October Full Moon to express the idea that it was the time of leaf-fall. Some typical names were 'Leaf-falling Moon', 'Falling Leaves Moon' or 'Fall Moon'. Some names, such as 'White frost on grass Moon' expressed the idea that significant frosts had arrived.

---

***Edmond Halley*** (1656–1742) is primarily known to the general public as an astronomer, and because of the famous comet named after him, the return of which he predicted, but did not live to see. (This was the very first such prediction.) However, he also made contributions to many different fields of science.

As far as meteorology is concerned, Halley made various voyages – he was commissioned into the Royal Navy – to investigate terrestrial magnetism, but these led to fundamental understanding of wind patterns. His voyage to St Helena in the South Atlantic (returning in May 1678) was of particular significance. From data he obtained on that voyage, he eventually published in 1686 a paper and a chart of the wind directions (particularly the trade winds and monsoons) around the world. This was of fundamental importance in establishing the details of the circulation of the atmosphere. He also found the important relationship between barometric pressure and height above sea level.

---

*Edmond Halley (1656–1742), after whom the Antarctic research station is named, in an undated painting by Thomas Murray.*

# Weather Extremes

| Country | Temp. | Location | Date |
| --- | --- | --- | --- |
| *Maximum temperature* | | | |
| England | 29.4°C | March (Cambridgeshire) | 1 Oct. 1985 |
| Northern Ireland | 24.1°C | Strabane (Co. Tyrone) | 10 Oct. 1969 |
| Scotland | 27.4°C | Tillypronie (Aberdeenshire) | 3 Oct. 1908 |
| Wales | 28.2°C | Hawarden Bridge (Flintshire) | 1 Oct. 2011 |
| *Minimum temperature* | | | |
| England | -10.6°C | Wark (Northumberland) | 17 Oct. 1993 |
| Northern Ireland | -7.2°C | Lough Navar Forest (Co. Fermanagh) | 18 Oct. 1993 |
| Scotland | -11.7°C | Dalwhinnie (Inverness-shire) | 28 Oct. 1948 |
| Wales | -9.0°C | St Harmon (Powys) | 29 Oct. 1983 |

| Country | Pressure | Location | Date |
| --- | --- | --- | --- |
| *Maximum pressure* | | | |
| Scotland | 1045.6 hPa | Dyce (Aberdeenshire) | 31 Oct. 1956 |
| *Minimum pressure* | | | |
| Scotland | 946.8 hPa | Cawdor Castle (Nairn) | 14 Oct. 1891 |

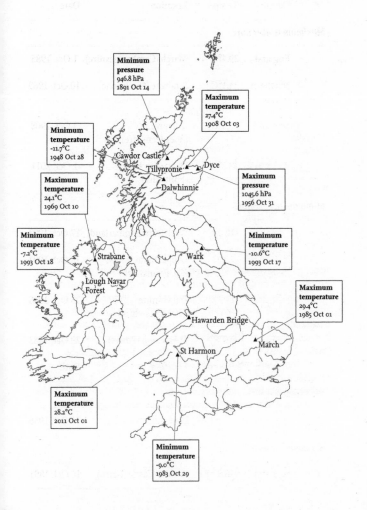

**Minimum pressure**
946.8 hPa
1891 Oct 14

**Maximum temperature**
27.4°C
1908 Oct 03

**Minimum temperature**
-11.7°C
1948 Oct 28

Cawdor Castle
Tillypronie    Dyce
Dalwhinnie

**Maximum pressure**
1045.6 hPa
1956 Oct 31

**Maximum temperature**
24.1°C
1969 Oct 10

**Minimum temperature**
-7.2°C
1993 Oct 18

Strabane

Wark

**Minimum temperature**
-10.6°C
1993 Oct 17

Lough Navar
Forest

**Maximum temperature**
29.4°C
1985 Oct 01

Hawarden Bridge

March

St Harmon

**Maximum temperature**
28.2°C
2011 Oct 01

**Minimum temperature**
-9.0°C
1983 Oct 29

167

# The Weather in October 2021

| Observation | Location | Date |
| --- | --- | --- |
| *Max. temperature*<br>22.9°C | Thornes Park<br>(West Yorkshire) | 8 October |
| *Min. temperature*<br>-3.6°C | Redesdale Camp<br>(Northumberland) | 16 October |
| *Rainfall*<br>22.6 mm | Honister Pass<br>(Cumbria) | 28 October |
| *Wind gust*<br>33 m/s<br>(65 knots or 75 mph) | Needles Old Battery<br>(Isle of Wight) | 31 October |

Most of the month had unsettled weather, with some extreme events at the end of the month, although there was a quieter spell during the second and part of the third weeks of the month. October opened with strong winds in the north of Scotland, and both winds and rain affected south Wales, with homes flooded in Rhondda Cynon Taf. Flooding also affected parts of London on October 5. There was prolonged rain in western Scotland during the second week of the month. Overall, southern Scotland, Cumbria, the northwest of Wales and southern England were extremely wet, with some locations receiving 50 per cent more rain than in an average October.

There was even twice the usual amount of rain at places in Cumbria.

The month's maximum temperature of 22.9°C was recorded at Thornes Park in West Yorkshire on October 8, during the quieter spell of weather that occurred at that time. The temperature fell to -3.6°C at Redesdale Camp in Northumberland on October 16.

The second half of October was very unsettled. There was flooding in Glamorgan and Carmarthenshire on October 20. A possible tornado caused damage on Merseyside and in Greater Manchester. The south of England was hit by very heavy rain on the night of October 20/21, with severe flooding in Essex. There was widespread flooding of roads in the London area, and even farther afield in Huntingdon and Southend. There was considerable disruption to both train and bus services. The floods extended farther west, with flooding in West Sussex, generally along the south coast as far as Devon and Cornwall, and in Somerset.

The last week of the month saw further problems. There was widespread flooding in southern Scotland and Cumbria, with 22.6 mm of rain at Honister Pass in the 24 hours to 9:00 am on October 28. In Wales, rail services were disrupted in both southern and northern areas of the country by flooding and severe problems were encountered on the roads. There was similar disruption in southern England, from Oxfordshire southwards, with flooding in the London area, northern Hampshire and Surrey. It became very windy and chilly during the last day of the month (October 31) with gusts of 33 m/s (65 knots or 75 mph) recorded at the Needles Old Battery on the Isle of Wight. The winds and cool weather spread to the north into Wales, with some thundery showers.

# Sunrise and Sunset 2023

| Location | Date | Rise | Azimuth ° | Set | Azimuth ° |
|----------|------|------|-----------|-----|-----------|
| **Belfast** | | | | | |
| | Oct 01 (Sun) | 06:25 | 94 | 18:01 | 265 |
| | Oct 11 (Wed) | 06:44 | 101 | 17:36 | 259 |
| | Oct 21 (Sat) | 07:04 | 107 | 17:12 | 252 |
| | Oct 31 (Tue) | 07:24 | 113 | 16:50 | 246 |
| **Cardiff** | | | | | |
| | Oct 01 (Sun) | 06:13 | 94 | 17:51 | 266 |
| | Oct 11 (Wed) | 06:29 | 100 | 17:29 | 260 |
| | Oct 21 (Sat) | 06:46 | 106 | 17:08 | 254 |
| | Oct 31 (Tue) | 07:04 | 112 | 16:48 | 248 |
| **Edinburgh** | | | | | |
| | Oct 01 (Sun) | 06:15 | 94 | 17:49 | 265 |
| | Oct 11 (Wed) | 06:35 | 101 | 17:23 | 258 |
| | Oct 21 (Sat) | 06:56 | 108 | 16:58 | 252 |
| | Oct 31 (Tue) | 07:17 | 114 | 16:35 | 245 |
| **London** | | | | | |
| | Oct 01 (Sun) | 06:01 | 94 | 17:39 | 266 |
| | Oct 11 (Wed) | 06:18 | 100 | 17:17 | 260 |
| | Oct 21 (Sat) | 06:35 | 104 | 16:56 | 254 |
| | Oct 31 (Tue) | 06:53 | 112 | 16:36 | 248 |

*Note that all times are in Universal Time (UT), otherwise known as Greenwich Mean Time (GMT). These times do not take Summer Time (BST) into account.*

# Moonrise and Moonset 2023

| Location | Date | Rise | Azimuth ° | Set | Azimuth ° |
|---|---|---|---|---|---|
| **Belfast** | | | | | |
| | Oct 01 (Sun) | 18:37 | 61 | 09:14 | 294 |
| | Oct 11 (Wed) | 02:36 | 68 | 17:04 | 286 |
| | Oct 21 (Sat) | 15:05 | 142 | 21:16 | 220 |
| | Oct 31 (Tue) | 17:39 | 40 | 11:14 | 319 |
| **Cardiff** | | | | | |
| | Oct 01 (Sun) | 18:37 | 64 | 08:52 | 292 |
| | Oct 11 (Wed) | 02:32 | 70 | 16:47 | 285 |
| | Oct 21 (Sat) | 14:28 | 137 | 21:29 | 224 |
| | Oct 31 (Tue) | 17:52 | 44 | 10:39 | 314 |
| **Edinburgh** | | | | | |
| | Oct 01 (Sun) | 18:21 | 60 | 09:06 | 295 |
| | Oct 11 (Wed) | 02:20 | 67 | 16:56 | 287 |
| | Oct 21 (Sat) | 15:06 | 144 | 20:52 | 217 |
| | Oct 31 (Tue) | 17:16 | 37 | 11:14 | 321 |
| **London** | | | | | |
| | Oct 01 (Sun) | 18:24 | 64 | 08:40 | 292 |
| | Oct 11 (Wed) | 02:20 | 70 | 16:36 | 285 |
| | Oct 21 (Sat) | 14:17 | 137 | 21:16 | 224 |
| | Oct 31 (Tue) | 17:39 | 44 | 10:27 | 314 |

*Note that all times are in Universal Time (UT), otherwise known as Greenwich Mean Time (GMT). These times do not take Summer Time (BST) into account.*

# Twilight Diagrams 2023

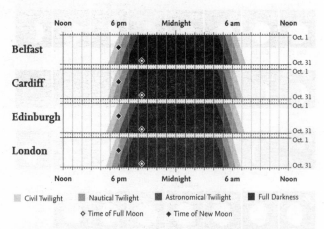

The exact times of the Moon's major phases are shown on the diagrams opposite.

---

*Rear-Admiral Sir Francis Beaufort* (1774–1857) was a British naval officer who in 1806 devised a means of estimating wind strength at sea. His scheme was not adopted by the Royal Navy until 1838. Beaufort's scale (including an adaptation for use on land) is still in use today (pages 248–251).

In 1829, Beaufort was appointed head of the Admiralty's Hydrographic Office, a post that he held for 25 years. Under his leadership, the Office became the world's leading hydrographic organisation. Beaufort made major contributions to many scientific fields, including geography, geodesy, oceanography, and astronomy, as well as meteorology.

---

# The Moon's Phases and Ages 2023

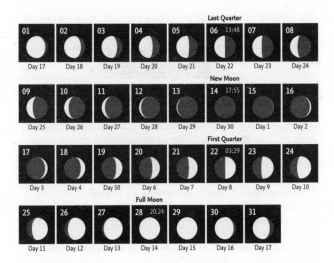

**Last Quarter**

| 01 | 02 | 03 | 04 | 05 | 06 13:48 | 07 | 08 |
|---|---|---|---|---|---|---|---|
| Day 17 | Day 18 | Day 19 | Day 20 | Day 21 | Day 22 | Day 23 | Day 24 |

**New Moon**

| 09 | 10 | 11 | 12 | 13 | 14 17:55 | 15 | 16 |
|---|---|---|---|---|---|---|---|
| Day 25 | Day 26 | Day 27 | Day 28 | Day 29 | Day 30 | Day 1 | Day 2 |

**First Quarter**

| 17 | 18 | 19 | 20 | 21 | 22 03:29 | 23 | 24 |
|---|---|---|---|---|---|---|---|
| Day 3 | Day 4 | Day 50 | Day 6 | Day 7 | Day 8 | Day 9 | Day 10 |

**Full Moon**

| 25 | 26 | 27 | 28 20:24 | 29 | 30 | 31 |
|---|---|---|---|---|---|---|
| Day 11 | Day 12 | Day 13 | Day 14 | Day 15 | Day 16 | Day 17 |

*Anticyclone*
A high-pressure area. Winds circulate around anticyclones in a clockwise direction in the northern hemisphere. (Anticlockwise in the southern hemisphere.) Anticyclones are slow-moving systems (unlike depressions) and tend to extend their influence slowly from an existing centre.

# October – In this month

**1 October 1956** – The RAF's very first Avro delta-winged Vulcan nuclear bomber crashed in a very severe thunderstorm and high winds on its final approach to Heathrow airport. Two of the six people on board survived.

**3 October 1066** – A southerly wind, after weeks of northerly storms and heavy seas, allows William of Normandy, with his invasion fleet, to make an unopposed landing in East Sussex.

**7 October 1869** – A great storm and flood were predicted by Lieutenant Saxby, a naval engineer and 'lunarian'. Saxby predicted what he termed 'extreme atmospheric disturbances' and a major storm as a result of the alignment of the Moon with the equator and other astronomical events. He warns that 'all are in danger of engulfment by wind or wave'. There was great expectation and thousands of people lined the shore and river embankments, expecting a storm and flooding. 'Saxby's Flood' did not occur, there was just a high spring tide. The weather was calm with a gentle breeze.

**11 October 1957** – The Windscale plant in Cumbria, producing plutonium for nuclear bombs, caught fire and radioactive iodine was released into the atmosphere. Initial reports were that the radioactive cloud was blown out to sea towards the Isle of Man. This was later denied. The Low Level Radiation Campaign accused the government of 'covering up' the situation, with record sheets removed from the Winscale meteorological record. The true situation (particularly the force and direction of the wind) remains unknown to this day.

**17 October 1091** – The very first recorded British tornado destroyed the wooden London Bridge and many houses (see page 90).

**19 October 1694** – An intense storm in Culbin, Moray, Scotland began to shift sand dunes to such an extent that fertile farmland became covered in sand. The shifting sand eventually buried houses and the laird's mansion. The Culbin Sands were later planted with a pine forest.

**21 October 1966** – A coal tip, partly mobilised by heavy rain, collapsed and buried the Pantglas Junior School school at Aberfan, a village near Merthyr Tydfil (see pages 174 and 175), with 144 fatalities, 116 of which were children at the school.

**24 October 1597** – A north-easterly storm prevented the so-called 'Third Armada' from making landfall on the shores of England. Following the great Armada of 1588, the country is essentially undefended. Only the weather acts to protect the country.

**27 October 1913** – A tornado, 200 metres wide cut a path from eastern Devon to Lancashire. Edwardsville, near Merthyr Tydfil in Wales sees the most destruction of property, with six people killed (the greatest death toll from any British tornado) and many injured.

# The Aberfan Disaster

The Aberfan Disaster of 21 October 1966 was partly weather-related. Heavy rain combined with water from the underground spring destabilised the tip, causing it to collapse. At 9.15 am, Coal Tip No.7 collapsed and buried Pantglas Junior School. Some half a million tonnes of coal waste (slag) destroyed the school and several neighbouring houses. Of the 144 victims, 116 were children, and of the adults who died, five were teachers at the school.

The tip was the responsibility of the National Coal Board (NCB), and the waste tip had been established over an underground spring, in direct contravention of the NCB policy. The NCB initially claimed that no one knew of the existence of any springs under the tip.

On 25 October 1966, Parliament set up a tribunal to investigate the disaster. This eventually reported and on 28 April 1967, the tribunal reported that blame for the disaster lay on the National Coal Board and its local managers. Neither the NCB nor any of its staff were prosecuted.

There were a number of waste tips, some particularly dangerous, still surrounding the village of Aberfan. After an extensive campaign by the residents, these were removed, but not before the government forcibly took £150,000 from the fund (the Aberfan Disaster Fund), contributed by members of the public. This sum was eventually repaid many years later (in 1997) by the Welsh Government.

*Typhoon*
The term used for a tropical cyclone in the western Pacific Ocean. Typhoons are some of the strongest systems encountered anywhere on Earth.

*An aerial photograph, showing the spoil tips before Tip No.7 slipped. Comparison with the image below shows the extent of later tipping alongside Tip No.7.*

*This aerial photograph shows how Tip No.7 sent a flood of dark slurry over the school, overwhelming it and several nearby houses.*

November

# Introduction

The weather in the early part of November tends to resemble that in October, so it is definitely an autumnal month. However, morning mists and fogs are much more frequent and also become more persistent, with the decreasing power of the Sun that would otherwise 'burn the fogs off' during the day. Such long-lasting mists and fogs are particularly frequent in the English Midlands and arise either though radiation at night or through the advection of relatively warm, saturated air over cold ground. When such conditions have brought saturated air in an airflow from the west, the descent of the air over the Scottish highlands may result in much warmer temperatures in north-eastern Scotland, particularly in the area around Aberdeen.

Because there is often moderately high pressure over the near Continent at this time of the year, it tends to divert depressions from an easterly track. They then generally veer towards the north, running up the west coast of Britain. The result is heavy rain and strong winds in the west of the country and particularly heavy rainfall (orographic rain) on the mountains of Wales and Scotland. The south and east of England is often much drier

The Full Moon that fell in November, unlike the September 'Harvest Moon' and the October 'Hunter's Moon', did not, in European tradition, have a well-known, universally applied name. It was sometimes known as the 'Frosty Moon' in recognition of the fact that the weather was getting colder, moving towards full winter. On a few occasions we find it called the 'Oak Moon', although this title is more properly applied to the Full Moon in December. Sometimes, if the Full Moon in November was the very last before the winter solstice in December, it was known as the 'Mourning Moon'.

### *Early winter – November 20 to January 19*

This season typically sees an alternation between long periods of mild westerly weather and drier anticyclonic conditions. Roughly half of the years see this pattern, usually with the westerly episodes being wet and windy, with a succession of depressions arriving from the Atlantic. Short cold periods, generally lasting less than a week, occur in between the westerly episodes. Very cold conditions rarely arrive before the very end of January and generally become established in early February.

# Weather Extremes

| Country | Temp. | Location | Date |
|---|---|---|---|
| **Maximum temperature** | | | |
| England | 21.1°C | Chelmsford (Essex) Clacton (Essex) Cambridge (Cambridgeshire) Mildenhall (Suffolk) | 5 Nov. 1938 |
| Northern Ireland | 18.5° | Murlough (Co. Down) | 3 Nov. 1979 1 Nov. 2007 10 Nov. 2015 |
| Scotland | 20.6°C | Edinburgh Royal Botanic Garden Liberton (Edinburgh) | 4 Nov. 1946 |
| Wales | 22.4°C | Trawsgoed (Ceredigion) | 1 Nov. 2015 |
| **Minimum temperature** | | | |
| England | -15.5°C | Wycliffe Hall (North Yorkshire) | 24 Nov. 1993 |
| Northern Ireland | -12.2°C | Lisburn (Co. Antrim) | 15 Nov. 1919 |
| Scotland | -23.3°C | Braemar (Aberdeenshire) | 14 Nov. 1919 |
| Wales | -18.0°C | Llysdinam (Powys) | 28 Nov. 2010 |

| Country | Pressure | Location | Date |
|---|---|---|---|
| **Maximum pressure** | | | |
| Scotland | 1046.7 hPa | Aviemore (Inverness-shire) | 10 Nov. 1999 |
| **Minimum pressure** | | | |
| Scotland | 939.7 hPa | Monach Lighthouse (Outer Hebrides) | 11 Nov. 1877 |

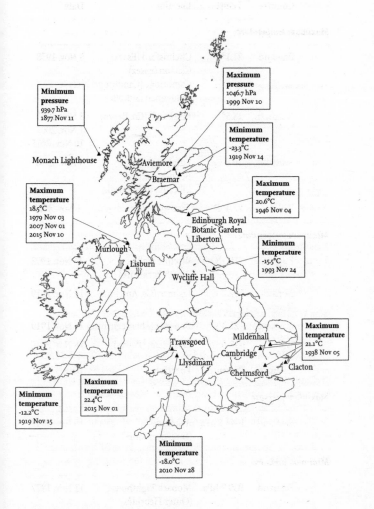

**Minimum pressure**
939.7 hPa
1877 Nov 11

Monach Lighthouse

**Maximum pressure**
1046.7 hPa
1999 Nov 10

**Minimum temperature**
-23.3°C
1919 Nov 14

Aviemore
Braemar

**Maximum temperature**
18.5°C
1979 Nov 03
2007 Nov 01
2015 Nov 10

**Maximum temperature**
20.6°C
1946 Nov 04

Edinburgh Royal
Botanic Garden
Liberton

Murlough

Lisburn

**Minimum temperature**
-15.5°C
1993 Nov 24

Wycliffe Hall

Trawsgoed
Mildenhall
Cambridge

**Maximum temperature**
21.1°C
1938 Nov 05

Llysdinam
Chelmsford
Clacton

**Minimum temperature**
-12.2°C
1919 Nov 15

**Maximum temperature**
22.4°C
2015 Nov 01

**Minimum temperature**
-18.0°C
2010 Nov 28

183

# The Weather in November 2021

| Observation | Location | Date |
| --- | --- | --- |
| *Max. temperature* 17.6°C | Nantwich, Reaseheath Hall (Cheshire) | 9 November |
| *Min. temperature* -8.7°C | Shap (Cumbria) Cromdale (Morayshire) | 29 November |
| *Rainfall* 67.2 mm | Achfary (Sutherland) | 19 November |
| *Wind gust* 44 m/s (85 knots or 98 mph) | Brizlee Wood (Northumberland) | 26 November |
| *Snow depth* 18 cm | Middleton, Hillside (Derbyshire) | 27 November |

The very beginning of November 2021 was unsettled, and during most of the month the weather was fairly quiet, until the very last few days. The month began with flooding in the Merseyside, Lancashire and Greater Manchester areas. The was also flooding in County Antrim and County Don in Northern Ireland. Strong winds affected northeast Scotland and there was some disruption to road traffic on November 7. Heavy rain affected some parts of Scotland on November 19 with Achfary in Sutherland recording 67.2 mm of rain in 24 hours on November 19.

Little occurred for the rest of the month until November 26 and 27 when the Met Office issued a red warning of winds from Storm Arwen, the first named storm of the season, likely to affect eastern Scotland and northern England. Fallen trees brought down by the high winds caused fatalities in County Antrim, Aberdeenshire and Cumbria. Both road and rail disruptions occurred in Scotland. There were also a number of power outages and many thousands of consumers were left without electricity. A large number of trees were uprooted in a forest near Kinbuck in Stirlingshire. Northern Ireland also experienced both road and rail disruptions.

Northern England was badly affected by Storm Arwen. There was considerable damage to properties, particularly in towns on the east coast. Both north-western and north-eastern counties suffered major power losses, affecting hundreds of thousands of consumers, especially in north-eastern areas. Trans-Pennine road routes were closed by snowfall. There were speed restrictions on railway lines all across the region, and many roads were blocked by fallen trees.

In England the end of the month was noticeably colder with widespread frosts. Even south-western England did not escape the effects of Storm Arwen. Winds increased over the whole country on November 26 as Storm Arwen moved out over the North Sea. On November 27, tens of thousands of customers in south-west England lost electrical power, and many roads were blocked. Maximum gusts of 44 m/s (98 mph) were recorded at Brizlee Wood (Northumberland) on November 26. Ice warnings in the last three days of the month hampered rescue efforts in the south-west. There were isolated sleet and snow showers in eastern coastal regions on November 28, but an area of snow moved south, producing a temperature of -8.7°C at Shap in Cumbria and a depth of lying snow of 18 cm at Middleton, Hillside (Derbyshire) on November 27.

The weather in Wales largely resembled that in England, starting mild, but turning colder and windy towards the end of the month (particularly on November 27). Snow moved into the region on November 28, although it then became milder.

# Sunrise and Sunset 2023

| Location | Date | Rise | Azimuth ° | Set | Azimuth ° |
|----------|------|------|-----------|-----|-----------|
| **Belfast** | | | | | |
| | Nov 01 (Wed) | 07:26 | 114 | 16:48 | 246 |
| | Nov 11 (Sat) | 07:46 | 120 | 16:29 | 240 |
| | Nov 21 (Tue) | 08:05 | 125 | 16:14 | 235 |
| | Nov 30 (Thu) | 08:20 | 128 | 16:04 | 232 |
| **Cardiff** | | | | | |
| | Nov 01 (Wed) | 07:06 | 112 | 16:46 | 247 |
| | Nov 11 (Sat) | 07:23 | 117 | 16:30 | 242 |
| | Nov 21 (Tue) | 07:40 | 122 | 16:16 | 238 |
| | Nov 30 (Thu) | 07:54 | 125 | 16:08 | 235 |
| **Edinburgh** | | | | | |
| | Nov 01 (Wed) | 07:19 | 115 | 16:33 | 245 |
| | Nov 11 (Sat) | 07:40 | 121 | 16:13 | 239 |
| | Nov 21 (Tue) | 08:00 | 126 | 15:56 | 234 |
| | Nov 30 (Thu) | 08:17 | 129 | 15:45 | 230 |
| **London** | | | | | |
| | Nov 01 (Wed) | 06:54 | 112 | 16:34 | 247 |
| | Nov 11 (Sat) | 07:12 | 119 | 16:17 | 242 |
| | Nov 21 (Tue) | 07:29 | 122 | 16:04 | 238 |
| | Nov 30 (Thu) | 07:43 | 125 | 15:56 | 235 |

*Note that all times are in Universal Time (UT), otherwise known as Greenwich Mean Time (GMT). These times do not take Summer Time (BST) into account.*

# Moonrise and Moonset 2023

| Location | Date | Rise | Azimuth ° | Set | Azimuth ° |
|----------|------|------|-----------|-----|-----------|
| **Belfast** | | | | | |
| | Nov 01 (Wed) | 18:19 | 36 | 12:31 | 324 |
| | Nov 11 (Sat) | 05:20 | 105 | 15:41 | 250 |
| | Nov 21 (Tue) | 14:24 | 107 | – | – |
| | Nov 30 (Thu) | 18:06 | 38 | 12:05 | 323 |
| **Cardiff** | | | | | |
| | Nov 01 (Wed) | 18:35 | 41 | 11:53 | 318 |
| | Nov 11 (Sat) | 05:03 | 104 | 15:37 | 252 |
| | Nov 21 (Tue) | 14:07 | 105 | – | – |
| | Nov 30 (Thu) | 18:20 | 43 | 11:28 | 318 |
| **Edinburgh** | | | | | |
| | Nov 01 (Wed) | 17:54 | 33 | 12:3 | 326 |
| | Nov 11 (Sat) | 05:11 | 106 | 15:27 | 250 |
| | Nov 21 (Tue) | 14:16 | 107 | – | – |
| | Nov 30 (Thu) | 17:42 | 36 | 12:07 | 325 |
| **London** | | | | | |
| | Nov 01 (Wed) | 18:21 | 41 | 11:42 | 319 |
| | Nov 11 (Sat) | 04:50 | 104 | 15:25 | 252 |
| | Nov 21 (Tue) | 13:55 | 106 | – | – |
| | Nov 30 (Thu) | 18:07 | 43 | 11:17 | 318 |

*Note that all times are in Universal Time (UT), otherwise known as Greenwich Mean Time (GMT). These times do not take Summer Time (BST) into account.*

# Twilight Diagrams 2023

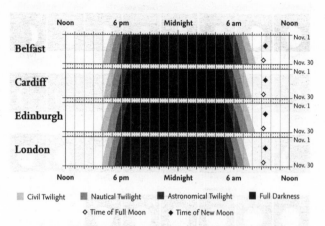

The exact times of the Moon's major phases are shown on the diagrams opposite.

## Sea breeze

A flow of air from the sea onto the land. The land heats more rapidly than the sea, so the air above it rises, drawing cooler air off the sea. There is a corresponding flow towards the sea at altitude. The air rises along a 'sea-breeze front', which may lie many kilometres inland, depending on the local geography.

# The Moon's Phases and Ages 2023

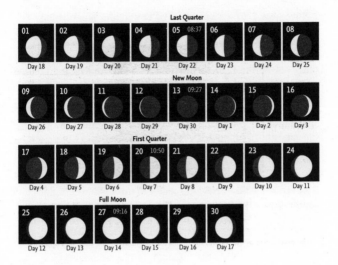

### Land breeze
A flow of air from the land towards the sea. At night, the land cools more quickly than the sea. The denser air flows out towards the sea. A 'land-breeze front' is sometimes marked (especially on satellite images) by a line of cumulus cloud, where the air rises to flow back towards the land at altitude.

# November – In this month

**2 November 1707** – A naval fleet of 21 warships is returning from the Mediterranean. Because of the weather and overcast skies they are navigating by dead reckoning. In effect they are lost and are hit by gales. The flagship, *HMS Association* and three other vessels the *Eagle*, *Romney* and *Firebrand* are wrecked, south-west of the Scilly Isles. Of 1673 on board, only 22 are saved. The disaster eventually leads to the passing of the Longitude Act, offering the prize of £20,000 for the determination of longitude at sea, then the formation of the Board of Longitude and the eventual award of the prize to Edward Harrison.

**5 November 2006** – Fireworks let off to mark Bonfire Night cause extremely heavy air pollution in London and in Lewes in Sussex (where there is a traditional Bonfire Night parade). The still and misty conditions serve to exacerbate the situation.

**6 November 1697** – Lightning, possibly ball lightning, strikes the armoury of Athlone Castle in Ireland. The resulting enormous explosion destroys most of the town.

**6 November 1889** – The famous rail bridge across the Firth of Forth is finally finished after eight years under construction. However, because of the cold weather, the two central spans do not meet. Only with warmer weather does the metal expand and the sections meet. The bridge is finally, officially opened in March 1890.

**8 November 1771** – After several days of continuous rain, Solway Moss, a peat upland near Carlisle, becomes saturated and slips to form a quagmire 1.6 km long. It smothers livestock, destroys buildings and inundates cottages, eventually covering hundreds of acres of farmland.

**15 November 1688** – William of Orange, who has been 'invited' to replace Janes II as King of England, made safe landfall at Brixham in Devon. His first attempt to sail to England was frustrated by a storm. The fleet initially made a navigational error and sailed past Torbay. However, the wind changes, and a 'Protestant wind' brings William of Orange safely to Brixham.

**19 November 1093** – The Battle of Alnwick took place after days of torrential rain. The River Aln is swollen and in flood. It prevents the retreat of the Scots, who are opposed to the Norman army. Malcolm III, King of Scotland, and his eldest son were killed in the battle.

**23 November 1824** – A storm surge and accompanying gale struck the south coast, particularly the coast of Dorset. It was later known as the 'Great Gale' and 'The Outrage'. High tide arrived early. The Chesil Bank was breached, and 80 houses behind it were destroyed. Some 50 or 60 people were drowned. Plymouth breakwater was destroyed as was the Cobb at Lyme Regis and the Promenade at Weymouth.

# TORRO and tornadoes in Britain

The Tornado and Storm Research Organisation (TORRO) was originally formed in 1974 as the Tornado Research Organisation just to investigate tornadoes, but its remit was later expanded (and the name changed) to include the investigation of severe storms, and other extreme weather. It now carries out research into electrical phenomena (such as ball lightning and extreme lightning), hail, blizzards and weather impacts. Tornado research, is, of course, still included. TORRO has developed scales for rating the severity of both tornadoes (see pages 252–255) and hail (see pages 256–257).

As mentioned previously (pages 90–91) tornadoes are often reported from Britain. Without detailed investigation or interviewing of responsible witnesses, it is difficult to derive an accurate intensity rating. Most tornadoes occur in England and Wales, and they are rare in Scotland and Northern Ireland. Tornadoes or waterspouts were reported on the day that the Tay Bridge collapsed (28 December 1879) and they may have acted to weaken the structure. At least two have been reported from Carrickfergus in County Antrim on 2 September 1775 and 12 June 1834.

The very first tornado recorded in Britain was the event that destroyed the old wooden London Bridge on 23 October 1091. It also destroyed some 600 buildings in London itself and damaged several churches. Some of the timbers of the church of St Mary le Bow, just under 8 metres long, were driven into the ground with such violence that little more than one metre remained visible. This tornado is nowadays regarded as one of the strongest (if not the strongest) tornado ever recorded in Britain. It is thought to have reached T8 on the TORRO Scale.

Another, extremely strong, tornado – again, perhaps the strongest – occurred on 23 October 1666. It tracked through four towns in Lincolnshire, uprooting trees, demolishing houses, and destroyed the church of Boothby Grafoe except part of the steeple. It has been rated T8–9 on the TORRO scale.

Another contender for the strongest British tornado was one that struck Southsea, a part of Portsmouth in Hampshire, on 14 December 1810. It had just a short recorded track, only given as from Old Portsmouth to Southsea Common (thus about 3 km), but caused immense damage, and, it is believed, no deaths. Some houses were completely destroyed and many others were so badly damaged that they had to be demolished; chimneys were blown down and the lead on a bank roof was rolled up and blown away.

There have been various outbreaks with more than one tornado reported on the same day. One such outbreak was on 3 January 1978, when some 10 to 14 tornadoes occurred on an active cold front that swept from north Humberside to Cambridgeshire. The tornadoes damaged houses, overturned caravans and uprooted trees. One, rated T5 or T6, passed south of Newmarket, damaging houses and a racing stable, and overturning cars.

At least three tornadoes, one particularly destructive, tracked from Wendover in Buckinghamshire to Linslade in Bedfordshire on 21 May 1950. They were accompanied by large hail and heavy rain.

Most British tornadoes are weak and relatively narrow. The widest appears to have been one near Fernhill Heath in Worcestershire on 22 October 1810. This seems to have had a width at the ground of 1.6 km (about 1 mile). It was very destructive and has been given a T8 rating.

Tornadoes are often associated with hailstorms. One of the most destructive was that which began at Horsey in Norfolk on 9 August 1843. It has been rated H7 on the TORRO hail scale (see pages 254–255). It left a wide 'hail swathe' right across the country to Stow-on-the-Wold in Gloucestershire. In places the hailstones were piled 1.5 metres deep. There was widespread destruction of glass, chimney pots and slates in the city of Cambridge.

December

# Introduction

For meteorologists, December is the beginning of winter. However, based on the actual weather that occurs during the month, it must be linked with late November and January, because, certainly nowadays, it sees very little severe weather. It may be – and often is – very windy with some notable storms during the month, but it is rarely very cold. The persistent idea of a white, snowy Christmas is actually a hangover from the writings of Charles Dickens in particular. In his young days, early in the nineteenth century, the weather around the time of Christmas was indeed more severe. Today, December is rarely as cold as January or February and on average, the month sees just two days when snow is lying.

December may see the winter solstice (December 22 in 2023), and the shortest day, but the temperature remains mild, mitigated by the sea, which is still relatively warm. As a whole then, December is marked by short days, accompanied by wet and windy weather. There have been some notably violent storms that have arrived in December, such as Storm

---

**Hadley cell**
One of the two atmospheric circulation cells that are driven by the hot air that rises in the tropics (i.e., along the equator or the heat equator that moves north or south with the seasons). The descending limbs are located over the sub-tropical highs at about latitudes 30° north and south.

---

Desmond and Storm Eva in December 2015. The rainfall from Storm Desmond broke the United Kingdom's 24-hour rainfall record, with 341.4 mm falling at Honister Pass, Cumbria, on December 5. Rain from Storm Eva added to the problems posed by Storm Desmond, and the precipitation caused severe flooding in Cumbia, where the towns of Appleby, Keswick and Kendal were all flooded on December 22. Some smaller locations were flooded three times during the month.

December is associated with Yule and Yuletide, originally a Germanic festival around the time of the winter solstice. The original pagan festival was taken over by the Christian church, and became Christmas. The Full Moon in December does not have a widely recognised name, unlike the Harvest and Hunter's Moons in September and October, but in the European tradition was sometimes known as 'the Moon before Yule' or even, occasionally, as the 'Wolf Moon', although that name is normally associated with the Full Moon of January.

---

*Ferrel cell*
One of the two atmospheric circulation cells that are the intermediate cells, between the Hadley cells, closest to the equator, and the polar cells in each hemisphere. The air, spreading out towards the poles from the sub-tropical highs is diverted towards the east by the rotation of the Earth, and forms the dominant westerlies that govern the weather in the middle latitudes.

---

# Weather Extremes

| Country | Temp. | Location | Date |
| --- | --- | --- | --- |
| *Maximum temperature* | | | |
| England | 17.7°C | Chivenor (Devon) | 2 Dec. 1985 |
| | | Penkridge (Staffordshire) | 11 Dec. 1994 |
| Northern Ireland | 16.0°C | Murlough (Co. Down) | 11 Dec. 1994 |
| Scotland | 18.3°C | Achnashellach (Highland) | 2 Dec. 1948 |
| Wales | 18.0°C | Aber (Gwynedd) | 18 Dec. 1972 |
| *Minimum temperature* | | | |
| England | -25.2°C | Shawbury (Shropshire) | 13 Dec. 1981 |
| Northern Ireland | -18.7°C | Castlederg (Co. Tyrone) | 24 Dec. 2010 |
| Scotland | -27.2°C | Altnaharra (Highland) | 30 Dec. 1995 |
| Wales | -22.7°C | Corwen (Denbighshire) | 13 Dec. 1981 |

| Country | Pressure | Location | Date |
| --- | --- | --- | --- |
| *Maximum pressure* | | | |
| Scotland | 1951.9 hPa | Wick (Caithness) | 24 Dec. 1926 |
| *Minimum pressure* | | | |
| Northern Ireland | 927.2 hPa | Belfast (Co. Antrim) | 8 Dec. 1886 |

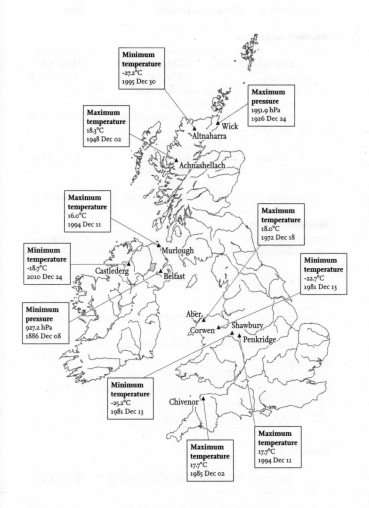

**Minimum temperature**
-27.2°C
1995 Dec 30

**Maximum pressure**
1951.9 hPa
1926 Dec 24

**Maximum temperature**
18.3°C
1948 Dec 02

Wick

Altnaharra

Achnashellach

**Maximum temperature**
16.0°C
1994 Dec 11

**Maximum temperature**
18.0°C
1972 Dec 18

Murlough

**Minimum temperature**
-18.7°C
2010 Dec 24

Castlederg

Belfast

**Minimum temperature**
-22.7°C
1981 Dec 13

**Minimum pressure**
927.2 hPa
1886 Dec 08

Aber.

Corwen

Shawbury

Penkridge

**Minimum temperature**
-25.2°C
1981 Dec 13

Chivenor

**Maximum temperature**
17.7°C
1985 Dec 02

**Maximum temperature**
17.7°C
1994 Dec 11

# The Weather in December 2021

| Observation | Location | Date |
| --- | --- | --- |
| **Max. temperature** 16.5°C | Bala (Gwynedd) | 31 December |
| **Min. temperature** -10.2°C | Braemar (Aberdeenshire) | 22 December |
| **Rainfall** 97.2 mm | Honister Pass (Cumbria) | 31 December |
| **Wind gust** 39 m/s (75 knots or 86 mph) | Aberdaron (Gwynedd) | 7 December |
| **Snow depth** 9 cm | Copley (Durham) | 26 December |

---

**Thermopause**
The transitional layer between the underlying mesosphere and the overlying exosphere. It is poorly defined and its altitude lies between 200 and 700 km, depending on solar activity. With strong solar activity, the altitude is depressed (i.e., it is lower).

---

Following Storm Arwen at the end of November, December began unsettled and also cold. There was heavy snowfall and road closures in Cumbria and Derbyshire on December 6. Storm Barra swept in the next day, December 7. It produced high winds over the whole country, but particularly in Northern Ireland, where there were power outages, and cancellations of local ferries as well as those to Cairnryan in Scotland. In Scotland itself there was heavy rain, snow over high ground and blustery showers. There was disruption to road traffic in England and Wales because of flooding and uprooted trees, together with isolated thunderstorms and snow on higher ground. The problems extended to East Anglia and south-east England around London with flooding and both road and rail disruptions. North-western England saw heavy snowfall on high ground. Wind gusts of 39 m/s (75 knots, 86 mph) occurred at Aberdaron in Gwynedd in Wales on December 7, and slightly slower gusts were recorded at the Needles Old Battery in the Isle of Wight on the same date.

A very quiet period of anticyclonic weather then ensued, and much later in the month on Boxing Day, December 26, snowfall affected roads across the Pennines and also over Honister Pass in the Lake District. Copley in Durham reported a lying snow depth of 9 cm on December 26, the maximum for the month. The month's minimum temperature was recorded the previous week with -10.2°C at Balmoral (Braemar) in Aberdeenshire on December 22.

On December 29, high winds and some heavy rain affected Northern Ireland and south-west England and Wales. In Northern Ireland, high winds and uprooted trees affected road travel. In south-west England there was road travel disruption by flooding and fallen trees. One of the Severn crossings was closed because of the high winds.

Heavy rain affected northern England, including at Honister Pass in Cumbria, where the month's maximum of 67.2 mm of rain was recorded in the 24 hours to 9.00 am on December 31 The heavy rain with some snow on higher ground in the north-west of England caused flooding and both road and rail travel disruption on December 30 and 31. There was heavy rain in the south-east and in East Anglia.

# Sunrise and Sunset 2023

| Location | Date | Rise | Azimuth ° | Set | Azimuth ° |
|---|---|---|---|---|---|
| **Belfast** | | | | | |
| | Dec 01 (Fri) | 08:22 | 128 | 16:03 | 232 |
| | Dec 11 (Mon) | 08:36 | 131 | 15:58 | 229 |
| | Dec 21 (Thu) | 08:44 | 132 | 15:59 | 228 |
| | Dec 31 (Sun) | 08:46 | 131 | 16:07 | 229 |
| **Cardiff** | | | | | |
| | Dec 01 (Fri) | 07:55 | 125 | 16:07 | 235 |
| | Dec 11 (Mon) | 08:08 | 127 | 16:04 | 232 |
| | Dec 21 (Thu) | 08:16 | 128 | 16:06 | 232 |
| | Dec 31 (Sun) | 08:18 | 128 | 16:13 | 232 |
| **Edinburgh** | | | | | |
| | Dec 01 (Fri) | 08:19 | 130 | 15:44 | 230 |
| | Dec 11 (Mon) | 08:33 | 132 | 15:39 | 227 |
| | Dec 21 (Thu) | 08:42 | 134 | 15:40 | 226 |
| | Dec 31 (Sun) | 08:44 | 133 | 15:48 | 227 |
| **London** | | | | | |
| | Dec 01 (Fri) | 07:44 | 125 | 15:55 | 235 |
| | Dec 11 (Mon) | 07:57 | 128 | 15:51 | 232 |
| | Dec 21 (Thu) | 08:05 | 128 | 15:53 | 232 |
| | Dec 31 (Sun) | 08:07 | 128 | 16:01 | 232 |

*Note that all times are in Universal Time (UT), otherwise known as Greenwich Mean Time (GMT). These times do not take Summer Time (BST) into account.*

# Moonrise and Moonset 2023

| Location | Date | Rise | Azimuth ° | Set | Azimuth ° |
|----------|------|------|-----------|-----|-----------|
| **Belfast** | | | | | |
| | Dec 01 (Fri) | 19:22 | 44 | 12:37 | 318 |
| | Dec 11 (Mon) | 07:07 | 131 | 14:27 | 226 |
| | Dec 21 (Thu) | 12:58 | 75 | 01:56 | 280 |
| | Dec 31 (Sun) | 20:58 | 67 | 11:28 | 297 |
| **Cardiff** | | | | | |
| | Dec 01 (Fri) | 19:32 | 48 | 12:04 | 314 |
| | Dec 11 (Mon) | 06:37 | 128 | 14:34 | 230 |
| | Dec 21 (Thu) | 12:52 | 76 | 01:41 | 279 |
| | Dec 31 (Sun) | 20:55 | 68 | 11:07 | 295 |
| **Edinburgh** | | | | | |
| | Dec 01 (Fri) | 19:00 | 42 | 12:37 | 320 |
| | Dec 11 (Mon) | 07:04 | 133 | 14:06 | 324 |
| | Dec 21 (Thu) | 12:45 | 75 | 01:46 | 280 |
| | Dec 31 (Sun) | 20:42 | 66 | 11:22 | 298 |
| **London** | | | | | |
| | Dec 01 (Fri) | 19:19 | 48 | 11:53 | 314 |
| | Dec 11 (Mon) | 06:26 | 128 | 14:22 | 230 |
| | Dec 21 (Thu) | 12:40 | 76 | 01:29 | 279 |
| | Dec 31 (Sun) | 20:42 | 68 | 10:55 | 295 |

*Note that all times are in Universal Time (UT), otherwise known as Greenwich Mean Time (GMT). These times do not take Summer Time (BST) into account.*

# Twilight Diagrams 2023

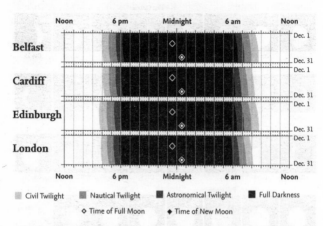

|  | Noon | 6 pm | Midnight | 6 am | Noon |

■ Civil Twilight    ■ Nautical Twilight    ■ Astronomical Twilight    ■ Full Darkness

◇ Time of Full Moon     ◆ Time of New Moon

**The exact times of the Moon's major phases are shown on the diagrams opposite.**

### *Polar cell*

One of the two cells, farthest from the equator, where cold air spreads out from the poles, heading in the general direction of the equator. The winds are easterly. This air meets the air in the Ferrel cells at the polar fronts, which tend to vary in latitude, and which are where the all-important depressions form.

# The Moon's Phases and Ages 2023

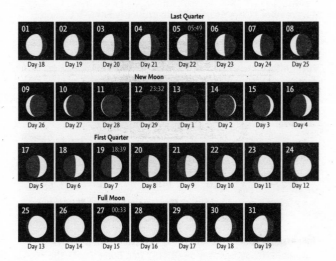

---

*Polar front*
The area where cold air from the poles meets warmer air that has spread out from the sub-tropical high-pressure regions. The conflict between the two types of air creates depressions which bring most of the changeable weather to countries in the middle latitudes.

---

# December – In this month

**2 December 1120** – Attempting to cross the English Channel, the *White Ship*, carrying Prince William Adelin, the only legitimate son and heir to King Henry I, was wrecked and the prince drowned. Despite marrying again, Henry I did not produce a male heir and named Matilda, his daughter, as successor. A nephew, Stephen, seized the throne in 1135, precipitating a civil war that lasted many years. Eventually, Matilda's son was crowned Henry II in 1155.

**4–5 December 1957** – The rainfall accompanying Storm Denis brought a new British record of 341.4 mm to Honister Pass in Cumbria (see page 206).

**9 December 1886** – The worst ever disaster struck the Royal National Lifeboat Institution (RNLI). In a Force 7 gale, the German vessel, *Mexico*, goes aground off Stockport. Three open, rowing, RNLI lifeboats are launched. One lifeboat succeeded in rescuing all 12 sailors. Of the other lifeboats, one, despite being self-righting, capsized with the loss of the crew of 14. The other lifeboat was washed ashore the next day with all 13 crew drowned.

**14 December 1810** – One of Britain's strongest tornadoes struck Portsmouth. There was major destruction along its short track (see page 191).

**22 December 1643** – During the English Civil War, a hard frost hardened usually muddy lanes, and enabled Sir William Waller to force-march his troops some 16 km, allowing him to surprise and rout the Royalist army at Alton in Hampshire.

**22 December 1991** – The giant wind turbines at Delabole in Cornwall, in the first wind farm ever established in Britain, began to operate and produce electricity.

**24 December 1841** – After days of rain, the earth of Sonning cutting on the Great Western Railway gave way in a landslip onto the tracks. A train plunged into the landslip, causing eight deaths and many injuries as passengers were thrown from the open carriages used at the time.

**26 December 1962** – Snow began to fall over England, driven by a bitterly easterly wind. It was the beginning of the phenomenal winter of 1962–63. Essentially all of the country saw snowfall over the next two months. The blanket of snow finally melts in March 1963.

# Seathwaite and Honister Pass

Two locations that are frequently mentioned in discussions of British rainfall records are those of Seathwaite and Honister Pass. Both are in the southern part of the Lake District in Cumbria.

Seathwaite is the wettest inhabited place in the British Isles. It is a small village in the Borrowdale valley and is at the foot of the valley on the B5289 road that snakes up the valley to Honister Pass and passes over the col into Buttermere.

In September 1996 more than 125 mm of rain fell on the village in one hour and the resulting floods caused great damage to an ancient packhorse bridge below the village. On 19–20 November 2009, the village recorded 314.4 mm of rain in the 24 hours to 9:00 am on November 20. This formed the record rainfall for any location in Britain until it was overtaken by the amount of 341.4 mm recorded at Honister Pass on 5 December 2015.

Honister Pass is a mountain pass between two valleys, that of Borrowdale on the south-west, and the southern end of Buttermere on the other side. The col, or saddle, at the watershed, with an altitude of 365 metres, is also known by the Cumbrian term of Honister Hause, and forms the natural boundary between Borrowdale and Buttermere.

The record rainfall at Honister Pass was brought by Storm Denis and it is suspected that the high rainfall was actually what is termed an 'atmospheric river'. A narrow stream of highly humid air that is dragged from the tropics ahead of the cold front of a depression.

*The top of Honister Pass as seen from the north-west side. The Honister Slate Mine and the Honister Hause Youth Hostel are not visible in this image.*

Additional Information

# Additional Information

# The Regional Climates of Britain

**1 South West England and the Channel Islands**

The south-western region may be taken to include Cornwall, Devon, Somerset, Gloucestershire, Dorset and the western portion of Wiltshire. This area is largely dominated by its proximity to the sea, although the northern and eastern portion of the region, being farther from the sea, often experiences rather different weather. In many respects the closeness of the Atlantic means that the weather resembles that encountered in the west of Ireland or the Hebrides. Generally, the climate is extremely mild, although that in the Scilly Isles is drier, sunnier and much milder than the closest part of the Cornish peninsula, just 40 km farther north. In the prevailing moist, south-westerly airstreams, the Scilly Isles are not only surrounded by the sea, but they are fairly flat with no hills to cause the air to rise and produce rain. The Channel Islands, by contrast, well to the east, are affected by their proximity to France and sometimes come under the influence of anticyclonic high-pressure conditions on the near Continent, so their overall climate tends to be more extreme.

Despite its generally mild weather, the region has experienced extremes, such as the exceptional snowfall in March 1891 that paralysed southern counties and introduced the word 'blizzard' to descriptions of British weather. The region also experienced the British rainstorm record of 279 mm in one single observational day (09.00 am one day to 08.59 am the next) occurring on 18 July 1955 at Martinstown in Dorset.

Most of the peninsula of Devon and Cornwall sees very few days of frost and some areas are almost completely frost-free. Temperatures are lower, of course, over the high ground of Bodmin Moor and Dartmoor. Indeed, those areas and the Mendip Hills and Blackdown Hills do all have slightly different climatic regimes. The influence of the Severn Estuary extends well inland, and actually has an affect on the weather in the Midlands (see page 218). In winter, it allows mild air to penetrate far inland. In Cornwall and Devon, particularly in summer, sea breezes from opposite sides of the peninsula converge over the high ground that runs along the centre of the peninsula, leading to the formation of major cumulonimbus

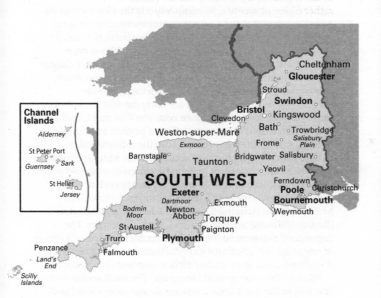

clouds and frequent showers, which may give extreme rainfall. It was this that led to the Lynmouth disaster in August 1952, when waters from a flash flood devastated the town and caused the deaths of 34 people. A somewhat similar situation arose in August 2004 in nearby Boscastle and Crackington Haven, although in that instance, no lives were lost.

The escarpment of the Cotswolds, overlooking the Severn valley, often proves to be a boundary between different types of weather. This is particularly the case when there is a north-westerly wind. Then heavy showers may affect areas on the high ground, while it is warmer and with less wind over the flatter land in the Severn valley and around Gloucester. It is often bitterly cold on the high ground above the escarpment.

## 2  South East England and East Anglia

The weather in the south-eastern corner of the country may be divided into two main areas: the counties along the south coast (Hampshire, West and East Sussex and Kent); and the Home Counties around London, and East Anglia, although East Anglia (Norfolk and Suffolk in particular) often experiences rather different conditions to the Home Counties.

The coastal strip from Hampshire (including the Isle of Wight) eastwards to southern Kent has long been recognised as the warmest and sunniest part of the British Isles. This largely arises from the longer duration of warm tropical air from the Continent when compared with the length of time that such air penetrates to more northern areas. The coastal strip from Norfolk to northern Kent does experience some warming effect when the winds are in the prevailing south-westerly direction, offsetting the cold effects produced by the North Sea. This coast may experience severe weather when there is an easterly or north-easterly airflow over the North Sea. This is particularly the case in winter: cold easterly winds bring significant snowfall to the region. It is also a feature of spring and early summer when temperatures are reduced when there is an onshore wind off the North Sea.

Frosts and frost hollows are a feature of the South and North Downs, the Chiltern Hills (in Berkshire, Bedfordshire and

Hertfordshire), and in the high ground in East Anglia. Here, the chalk subsoil loses heat by night as do the sandy soils of Surrey and Breckland in Norfolk, leading to ground frost in places in any month of the year.

Along the south coast there is a tendency for most rain to fall in the autumn and winter, whereas for the rest of the region (the Home Counties and East Anglia) precipitation tends to occur more-or-less equally throughout the year. Because of the reliance upon groundwater, the whole region sometimes suffers from drought, when winter rains (in particular) have been insufficient to recharge the underground reserves.

## 3 The Midlands

The Midlands region consists of a very large number of counties: Shropshire, Herefordshire, Worcestershire, Warwickshire, West Midlands, Staffordshire, Nottinghamshire, Lincolnshire, Leicestershire, Rutland, Northamptonshire and the southern part of Derbyshire (excluding the high ground of the High Peak in the north). Of all the regions of the British Isles, this is naturally the area which has the least maritime influence. The region has been likened to a shallow bowl surrounded by hills (the Welsh Marches, the Cotswolds, the Northamptonshire Escarpment, the Derbyshire Peak and the Staffordshire Moorlands) and with a slight dome in the centre (the Birmingham Plateau). In winter the warmest area is that closest to the Severn Valley, where warm south-westerly winds may penetrate inland, whereas in summer the warmest region lies to the north-east, farthest from those moderating winds. Yet the western area is also prone to very cold nights in autumn, winter and early spring. (The lowest temperature ever recorded in England was -26.1°C at Newport in Shropshire on 10 January 1982.) Frosts are a feature of the whole region, partly because of the sandy nature of most of the soils and also because of the lack of maritime influence.

Precipitation is fairly evenly spread across the region, although the west, along the Welsh border and the high northern area of the High Peak, experiences the highest rainfall. The east (Lincolnshire and the low ground in the

east of Nottinghamshire and Northamptonshire along the valleys of the Trent and Nene), tends to be drier. Because of the rain-shadow created by the Welsh mountains, over which considerable rainfall occurs, some areas of the west of this Midlands region are drier than might otherwise be expected. There is some increase in rainfall over the slightly higher ground of the Birmingham Plateau and also towards the south and the Cotswold hills. Towards East Anglia there is a strong tendency for most rain to occur in summer, when showers are most numerous. In the very hilly areas on the Welsh border and in the Peak District, the wettest months are December and January. The Derbyshire Peak and the Staffordshire Moorlands tend to experience considerable snowfall in winter, as do high areas of the Welsh Marches. On the lower ground to the east, snowfall is greater in the year than in the west. This is particularly the case when there are easterly or north-easterly winds that penetrate inland and bring snow from the North Sea.

## 4  North West England and the Isle of Man

The North West region consists of the land west of the Pennine chain, that is Cumbria and Lancashire in particular, especially including the mountainous Lake District in Cumbria, but also extends south to include Merseyside, Greater Manchester, Cheshire and the western side of Derbyshire. The region's weather tends to be mild and wetter in winter than regions to the east of the Pennines, and cooler in summer than regions to the south. It was, of course, the mild, relatively damp climate that was responsible for the region being the centre for the spinning of cotton, in contrast to the wool handled in the drier east. The maritime influence is seen in the fact that coastal areas are often warmer in winter and cooler in summer than areas farther inland.

There is a great difference in the amount of precipitation between the north and south of this region. The north, in Cumbria and the Lake District is notorious for the high rainfall. Seathwaite, in the Lake District, currently holds the record for rainfall in 24 hours and is the wettest inhabited location in Britain. The extreme rainfall has often contributed

to severe flooding, such as that in 2005 and 2009 in Carlisle, Cockermouth, Workington, Appleby and Keswick. By contrast, rainfall is much less over the Cheshire plain, which like much of Merseyside, actually lies in the rain shadow of the Welsh mountains and is thus much drier.

The prevailing wind from the south-west may give very high wind speeds over the high ground of the Pennines, while an easterly wind may produce the only named British wind, the viciously strong, and noisy, Helm wind, as air cascades over the escarpment west of Cross Fell and over the Eden valley in the north of Cumbria.

Most years see some early snow in autumn on the high fells, and in the north on the high ground it may be persistent, although it rarely lasts throughout the winter. The low ground along the coast and in the south sees relatively little snow and what does fall remains lying for just a few days.

## 5  North East England and Yorkshire

The region of North East England is well-defined by two geographical boundaries. On the west is the Pennine range, on the east, the North Sea. The northern boundary may be taken as the river Tweed and the southern as the estuary of the Humber. The region includes very high moorland in the north and west. In general the ground slopes down from the Pennine chain towards the east coast. There are, however, considerable areas of lower land, such as that in parts of Northumberland and, in particular, in the south of Yorkshire. Because the prevailing winds in Britain are from the west, they tend to deposit most of their rainfall over the Pennines, so that there is a rain shadow effect that reaches right across this region to the coast, and the whole region is drier than might otherwise be expected. With westerly winds the high fells also tend to break up the cloud cover, so that the whole area to the east is surprisingly sunny. On the other hand, the North Sea exerts a strong cooling effect, keeping general temperatures fairly low. The region is, however, open to easterly and northerly winds and tends to suffer from gales off the sea, accompanied by heavy rain or, in winter, by snow.

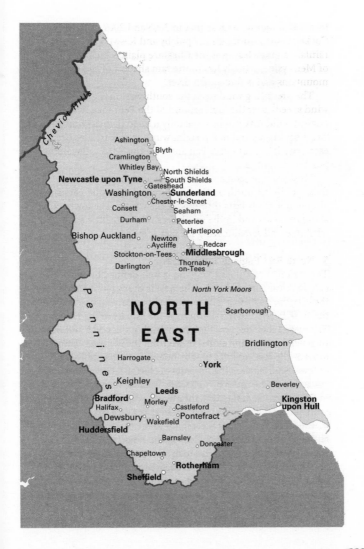

The North Sea never becomes particularly warm and exerts a chilling effect over the whole length of the region. Sea breezes, which occur because of the difference between the warm land and the cold sea, commonly set in, especially in late spring and often bring damp sea fog, or 'haar' to the coastal strip. Such conditions frequently arise when other parts of the country, especially the south-east, are enjoying settled, warm, anticyclonic weather. The cooling effect of the North Sea may be so great that there is a difference of as much as 10 degrees Celsius between the coastal strip and locations just a few kilometres inland. This difference tends to be greatest during the summer months.

Although the region has a low overall rainfall because of the effect of the Pennines to the west, when the wind is easterly it may produce prolonged heavy rainfall, as the air is forced to rise over the high ground, causing it to deposit its moisture as rain or, in winter, as snow. Long periods of heavy rain may occur from the easterly airflow on the northern side of depressions, the centres of which are tracking farther south across the country. Most of the rain is of this, frontal, type as there are few of the convective showers that produce heavy rainfall in regions farther south. Most of the rivers in the region have large catchment areas, extending well into the Pennines, and are therefore subject to episodes of flooding when there is prolonged rain. This is particularly the case in the south of the region, where the Ouse frequently floods around York and where the city of Hull often suffers.

## 6 Wales

Although Wales is being treated here as a single climatic region, in reality there are considerable differences in various areas. The whole country, is, of course, dominated by the long mountainous spine running from north to south. This area is not only high, but it is also exposed, windy and wet. To the west, the area around the whole of Cardigan Bay has a very maritime climate. It is windy, but may be particularly fine when

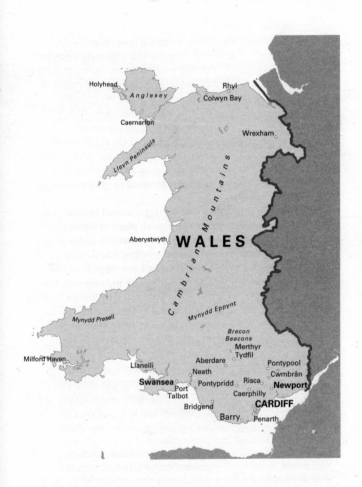

the mountains provide shelter from easterly winds. The driest area of Wales is in the northeast, in the lowlands bordering on Cheshire. Here there is a distinct rain shadow effect in the prevailing south-westerly maritime winds. To the south of this area, the country along the border with England is also subject to a rain shadow effect when the winds are westerly, but may experience considerable rainfall when depressions cross southern England.

The final distinct area is that of South Wales, from Pembrokeshire round to Glamorgan and southern Monmouthshire. This coastal area is generally mild, but tends to be wet and may be surprisingly cold in winter. The low-lying area around Cardiff, Newport and southern Monmouthshire is often affected in summer by fine weather over southern England, and may, at times, with southerly or south-westerly winds even derive some shelter from Exmoor on the other side of the Bristol Channel. In summer, warm conditions in southern England may also be transported west by easterly winds and affect the whole area. In winter, by contrast, the area may be cold as cold air drains out of the English Midlands along the Severn valley, bringing low stratus cloud or fog to the south-east of Wales.

## 7  Ireland

The climate of Ireland is dominated, as might be expected, by the Atlantic or, more specifically, by the warm waters of the North Atlantic Drift off its western coast. The whole island is subject to the maritime influence and is certainly far warmer than might otherwise be expected from its latitude. This is borne out by the extent, the quantity, and the quality of its grasslands. The length of time when the mean temperature exceeds 6°C, which is accepted as the limit for the growth of grass is exceptionally long. It is not for nothing that Ireland has earned the nickname of the 'Emerald Isle'. The climate is exceptionally equitable and there is very little variation in temperatures throughout the year. However, its location is firmly beneath one of the major storm tracks that are followed by depressions arriving from the Atlantic. By bearing the brunt of any severe storms, not only does the island act as an 'early

warning system' for the remainder of the British Isles, but also tends to temper their effects and reduce their severity for regions farther to the east.

Taken as a whole, the higher ground is located around the coasts of Ireland, with lower ground in the centre of the island. There is therefore a tendency for rainfall to be higher around the coasts than in the centre. One consequence of the warm sea is that in winter, in particular, the temperature difference between the cool land and the warm sea helps to strengthen developing depressions, so that frontal systems – and the rain that they bear – tend to be stronger in winter than in summer. Rainfall on hills close to the western coasts is greater at that time of the year, when there are also more showers, which add to the overall total amount of rain. This is particularly the case when unstable polar maritime air arrives in the wake of a depression. As it passes over the main flow of the North Atlantic Drift, it becomes strongly heated from below – sometimes by as much as 9 degrees Celsius – which thus increases its instability and the strength of the showers that are generated. All of which increases the rainfall.

As far as temperatures are concerned, there is very little variation across the island, although there tends to be a greater range in the north-east as compared with those prevailing over the south-west of Ireland. Distance from the coast also plays a part. The area with the greatest range of temperature is in southern Ulster, which experiences colder temperatures in late autumn and winter than areas along the southern and western coasts. From late spring until early autumn, maximum temperatures are higher in the north-east than in southern and western Ireland. In high summer, latitude does play a part, with temperatures being slightly higher in southern areas than along the north coast. One consequence of the equable maritime climate is that high temperatures, even in high summer, are rare, especially when compared with those that are experienced in southern England. On very rare occasions, in winter, the frigid air of the Siberian High may extend right across England and into Ireland and remains strong enough to overcome the influence of the warm air from the Atlantic. This was the case in

1962/1963 and in 2009/2010, but these winters were exceptional. Otherwise, frost is rare in coastal areas and occurs on just some 40 occasions in inland regions. As may be expected, snowfall is rare. Only in the extreme north-east does snow fall on an average of 30 days a year. In the far south-west, this figure decreases to about 5 days a year, and nowhere does snow lie for more than about a day or so.

The difference between the western coasts and the more sheltered inland areas is perhaps most obvious when wind strengths are compared. The strongest winds are observed on the northern coast of Ulster, which tends to be close to the area of the Atlantic over which depressions may undergo explosive deepening, with a consequent increase in wind speed. On rare occasions, such as the 'Night of the Big Wind', which began on the afternoon of 6 January 1839, highly destructive winds may extend right across the island, even into the far south.

## 8 Scotland

As with Wales, Scotland consists of several areas with diverse climates, which are a result of its mountainous as well as its maritime nature. Again as in the case with Wales, there are five distinct areas. Because they are at a distance from the mainland, the three island areas of the Hebrides (or Western Isles), Orkney and Shetland form one climatic area. In the west of the mainland, there is a distinction between the very mountainous Western Highlands region, running from Sutherland in the north, right the way down the west coast. The climate here is extremely wet because of the mountainous nature. To the east of this region, particularly in Caithness, Moray, and most of Aberdeenshire, although still a highland region (the Eastern Highlands), the area is shielded from the prevailing westerly winds by the Cairngorm Mountains and warmed by the föhn effect. It is, however, fully exposed to frigid northerly winds, especially in the north, in Caithness, and particularly so in winter. On the east coast farther to the south, the climate is essentially the same from the Moray Firth, down the coastal strip of Aberdeenshire, Angus, and Fife, across the Firth of Forth and as far as the eastern end of the Borders. Farther inland, the Central Lowlands and the western area of the Borders in Dumfries and Galloway and Ayrshire form yet another climatic region.

The highest temperatures are recorded in all areas in July (sometimes July and August in the outer islands), and there is a difference of some 2–3 degrees Celsius between the average temperature recorded in the Borders in the south and that found in the extreme north of the Eastern Highlands region (in Caithness). It is striking that the highest December and January temperatures in the whole of Britain have been recorded in northern Scotland. In both cases the temperature was 18.3°C, and both occurred as a result of the föhn effect in the lee of high mountains. On 2 December 1948, Ashnaschellach in the mountainous Western Highlands region experienced this temperature and on 26 January 2003 it was Aboyne in the Eastern Highlands region in Aberdeenshire. High temperatures tend to occur in the Western Highlands with southerly or south-easterly winds, whereas the highest in the Eastern Higlands occur with westerly winds. The lowest temperatures occur in December or January in all regions with the exception of the Hebrides, Orkney and Shetland, where the lowest temperatures are recorded in February. It is believed that lower temperatures than those recorded at Braemar on 11 February 1893 and again on 10 January 1982 and at Altnaharra on 30 December 1990 may have occurred at other locations, where there are no recording stations.

Scotland is known for its wet climate. There is, however, quite a striking difference between rainfall in the west (about 1250 mm per year in the outer islands and even more in the Western Highlands region), and rainfall in the eastern coastal region (that running from the Moray Firth down to the Borders). Here, yearly totals of just 650–750 mm are typical. This key feature of the Scottish climate is, of course, related to the mountainous nature of the land on the west, which receives most of the precipitation and shields the rest of the country. Precipitation also falls as snow and, once again, there is a distinct difference between the west and east of the country. Snowfall is also strongly dependent on altitude and here again

the Western Highlands region receives greater amounts of snow than elsewhere. In the Western Isles and particularly in western Ayrshire, days on which snow is lying are very few.

When it comes to sunshine, the eastern coastal strip is favoured, as well as the area of the western edge of the Borders in the south. The sunniest areas in Scotland are northern Fife, the northern shore of the Firth of Tay and the Mull of Galloway. Here, sunshine totals may even match those found in southern England.

# Clouds

*Altocumulus clouds.*

### Altocumulus
Altocumulus clouds (Ac) are medium-level clouds, with bases at 2–6 km, that, like all other varieties of cumulus, occur as individual, rounded masses. Although they may appear in small, isolated patches, they are normally part of extensive cloud sheets or layers and frequently form when gentle convection occurs within a layer of thin altostratus (page 234) breaking it up into separate heaps or rolls of cloud.

Altocumulus clouds may also take on the appearance of flat 'pancakes', but whatever the shape of the individual cloudlets, they always show some darker shading, unlike cirrocumulus. Blue sky is often visible between the separate masses of cloud – at least in the nearer parts of the layer.

Layers of altocumulus move as a whole, carried by the general wind at their height, but wind shear often causes the cloudlets to become arranged in long rolls or billows, which usually lie across the wind direction. High altocumulus or cirrocumulus of this type give rise to beautiful clouds that are commonly known as 'mackerel skies'.

### Altostratus

Altostratus (As) is a dull, medium-level white or bluish-grey cloud in a relatively featureless layer, which may cover all or part of the sky. When illuminated by the rising or setting Sun, gentle undulations on the base may be seen, but these should not be confused with the regular ripples that often occur in altocumulus (page 233).

As with stratus, altocumulus may be created by gentle uplift. This frequently occurs at a warm front, where initial cirrostratus thickens and becomes altostratus, and the latter may lower and become rain-bearing nimbostratus (page 241). Patches or larger areas of altostratus may remain behind fronts, shower clouds or larger, organised storms. Conversely, altostratus may break up into altocumulus. Convection may then eat away at the cloud, until nothing is left.

*Altostratus.*

*Cirrus.*

## Cirrus

Cirrus (Ci) is a wispy, thread-like cloud that normally occurs high in the atmosphere. Usually white, it may seem grey when seen against the light if it is thick enough.

Cirrus consists of ice crystals that are falling from slightly denser heads where the crystals are forming. In most cases wind speeds are higher at upper levels, so the heads move rapidly across the sky, leaving long trails of ice crystals behind them. Occasionally, the crystals fall into a deep layer of air moving at a steady speed. This can produce long, vertical trails of cloud.

## Cirrocumulus

Cirrocumulus (Cc) is a very high white or bluish-white cloud, consisting of numerous tiny tufts or ripples, occurring in patches or larger layers that may cover a large part of the sky. The individual cloud elements are less than 1° across. They are sometime accompanied by fallstreaks of falling ice crystals. Unlike altocumulus (page 233) the small cloud elements do not show any shading. They are outlined by darker regions where the clouds are very thin or completely missing.

*Cirrocumulus.*

The cloud layer is often broken up into a regular pattern of ripples and billows. Clouds of this sort are often called a 'mackerel sky', although the term is sometimes applied to fine, rippled altocumulus. In fact, the differences between cirrocumulus and altocumulus are really caused only because the latter are lower and thus closer to the observer.

### Cirrostratus

Cirrostratus (Cs) is a thin sheet of ice-crystal cloud and is most commonly observed ahead of the warm front of an approaching depression, when it is often the second cloud type to be noticed, after individual cirrus streaks. On many occasions, however, cirrostratus occurs as such a thin veil that it goes unnoticed, at least initially, until one becomes aware that the sky has lost its deep blue colour and has taken on a slightly milky appearance. The Sun remains clearly visible (and blindingly bright) through the cloud, but as the cirrostratus thickens a slight drop in temperature may become apparent.

*A solar halo in thin, almost invisible, cirrostratus cloud.*

Once you realise that cirrostratus is present, it pays to check the sky frequently, because as it thickens it may display striking halo phenomena. This stage often passes fairly quickly as the cloud continues to thicken and lower towards the surface, eventually turning into thin altostratus (page 234).

Cirrostratus often has a fibrous appearance, especially if it arises from the gradual increase and thickening of individual cirrus streaks. Because the cloud is so thin, and contrast is low, the fibrous nature is easier to see when the Sun is hidden by lower, denser clouds or behind some other object.

*Cumulus.*

## Cumulus

Cumulus clouds (Cu) are easy to recognise. They are the fluffy clouds that float across the sky on a fine day, and are often known as 'fair-weather clouds'. The individual heaps of cloud are generally well separated from one another – at least in their early stages. They have rounded tops and flat, darker bases. It is normally possible to see that these bases are all at one level. Together with stratus (page 241) and stratocumulus (page 242) they form closer to the ground than other cloud types.

The colour of cumulus clouds, like that of most other clouds, depends on how they are relative to the Sun and the observer. When illuminated by full sunlight, they are white – often blindingly white – but when seen against the Sun, unless they are very thin, they are various shades of grey.

## Cumulonimbus

Cumulonimbus (Cb) is the largest and most energetic of the cumulus family. It appears as a vast mass of heavy, dense-looking cloud that normally reaches high into the sky. Its upper portion is usually brilliantly white in the sunshine, whereas its lower portions are very dark grey. Unlike the flat base of a cumulus, the bottom of a cumulonimbus is often ragged and it may even reach down to just above the ground. Shafts of precipitation are frequently clearly visible.

Cumulonimbus clouds consist of enormous numbers of individual convection cells, all growing rapidly up into the sky. Although cumulonimbus develop from tall cumulus, a critical difference is that at least part of their upper portions has changed from a hard 'cauliflower' appearance to a softer, more fibrous look. This is a sign that freezing has begun in the upper levels of the cloud.

*Cumulonimbus.*

*Nimbostratus.*

*Stratus.*

## Nimbostratus

Nimbostratus (Nb) is a heavy, dark grey cloud with a very ragged base. It is the main rain-bearing cloud in many frontal systems. Shafts of precipitation (rain, sleet or snow) are visible below the cloud, which is often accompanied by tattered shreds of cloud that hang just below the base.

Just as cirrostratus often thickens and grades imperceptibly into altostratus, so the latter may thicken into nimbostratus. Once rain actually begins, or shafts of precipitation are seen to reach the ground, it is safe to call the cloud nimbostratus.

## Stratus

Stratus (St) is grey, water-droplet cloud that usually has a fairly ragged base and top. It is always low and frequently shrouds the tops of buildings. Indeed it is identical to fog, which may be regarded as stratus at ground level. Although the cloud may be thin enough for the outline of the Sun to be seen clearly through it, in general it does not give rise to any optical phenomena. It forms under stable conditions and is one of the cloud types associated with 'anticyclonic gloom'.

Stratus may form either by gentle uplift (like the other stratiform clouds) or when nearly saturated air is carried by a gentle wind across a cold surface, which may be either land or sea. Normally low wind speeds favour its occurrence, because mixing is confined to a shallow layer near the ground. When there is a large temperature difference between the air and the surface, however, stratus may still occur, even with very strong winds. Stratus also commonly forms when a moist air-stream brings a thaw to a snow-covered surface.

There is very little precipitation from stratus, but it may produce a little drizzle or, when conditions are cold enough, even a few snow or ice grains. Ragged patches of stratus, called 'scud' by sailors, often form beneath rain clouds such as nimbostratus or cumulonimbus, especially where the humid air beneath the rain cloud is forced to rise slightly, such as when passing over low hills.

## Stratocumulus

Stratocumulus (Sc) is a low, grey or whitish sheet of cloud, but unlike stratus it has a definite structure. There are distinct, separate masses of cloud that may be in the form of individual clumps, broader 'pancakes', or rolls. Sometimes these may be defined by thinner (and thus whitish) regions of cloud, but frequently blue sky is clearly visible between the masses of cloud.

Stratocumulus indicates stable conditions, and only slow changes to the current weather. It generally arises in one of two ways: either from the spreading out of cumulus clouds that reach an inversion, or through the break-up of a layer of stratus cloud. In the first case, the tops of the cumulus flatten and spread out sideways when they reach the inversion, producing clouds that are fairly even in thickness, with flat tops and bases. Initially, perhaps early in the day, there may be large areas of clear air, but the individual elements gradually merge to cover a larger area, or even completely blanket the sky.

In the second case, shallow convection (whose onset is often difficult to predict) begins within a sheet of stratus, causing the layer to break up. The regions of thinner cloud or clear air indicate where the air is descending, and the thicker, darker centres where it is rising.

*Stratocumulus.*

## Cloud heights

The height of clouds is usually given in feet (often with approximate metric equivalents). This may seem odd, when all other details of clouds, and meteorology in general, uses metric (SI) units. It is, however, a hangover from the way in which aircraft heights are specified. When aviation became general between the two World Wars, most commercial flying took place in the United Kingdom and America, so heights of aircraft and airfields were given in feet. It was obviously essential for cloud heights to be the same. The practice has continued: the heights of airfields, aircraft and clouds are still given in feet. The World Meteorological Organization recognises three ranges of cloud heights: low, middle and high. Clouds are specified by the height of their bases, not by that of their tops. The three divisions are:

Low clouds (bases 6500 feet or lower, approx. 2 km and below): cumulus (page 238), stratocumulus (page 242), stratus (page 241).

Middle clouds (bases between 6500 and 20,000 feet, approx. 2 to 6 km): altocumulus (page 233), altostratus (page 234), nimbostratus (page 241).

High clouds (bases over 20,000 feet, above 6 km): cirrus (page 235), cirrocumulus (page 235), cirrostratus (page 236).

One cloud type, cumulonimbus (page 239) commonly stretches through all three height ranges. Nimbostratus, although nominally a middle-level cloud, is frequently very deep and although it has a low base, may extend to much higher altitudes.

# Cyclones

In meteorology, a cyclone is a system in which air rotates around a low-pressure centre. All cyclones have inwardly-spiralling winds. The direction of these winds is anticlockwise in the northern hemisphere and clockwise in the southern. There is a great difference between the extratropical cyclones (depressions) that affect the middle latitudes and the tropical cyclones that originate closer to the equator.

Tropical cyclones are known by different names, depending on where they occur on Earth. They are known as 'hurricanes' over the North Atlantic Ocean and the north-eastern Pacific Ocean; 'typhoons' over the western Pacific Ocean; and just 'tropical cyclones' or 'cyclones' over the Indian Ocean and the southern Pacific Ocean. Tropical cyclones are very rare over the South Atlantic. Despite the different names, they are all basically the same. They are generally born where the trade winds converge at what is known as the Intertropical Convergence Zone (ITCZ), which shifts north and south of the equator with the seasons. Hurricanes in the North Atlantic generally arise from what is known as an 'easterly wave' or 'tropical wave', a low-pressure zone that lies on the equatorial side of the ITCZ and moves westwards across the Atlantic.

Tropical cyclones arise where the sea surface temperature is high. (It is generally accepted that the sea surface temperature must be at least 26–27°C.) This produces high humidity, rapidly rising air and cloud, giving rise to heavy showers and then thunderstorms. The intensity of tropical cyclones largely depends on the heat content of the oceanic waters and how rapidly the cyclone passes over that region.

The very high humidity in the surface air, and the vigorous resulting convection creates strong showers and then thunderstorms. These tend to clump together and set up a common circulation. The direction in which the circulation occurs is determined by the Coriolis Force, which is minimum at the equator, but increases towards the poles. For this reason tropical cyclones do not arise over the equator, but approximately 5° north or south of it (near the ITCZ). In the north, the Coriolis Force and surface friction force the circulation to be anticlockwise at the surface. (Clockwise in the southern hemisphere.)

The groups of thunderstorms cluster together to create high towers of cloud and high winds that encircle the low-pressure core. The clustered thunderstorms create spiral bands of heavy rain around the centre. Rather than blowing into a central point, the strong winds in tropical cyclones form a closed circulation in the form of a ring around a calm area at the centre. On satellite images it is often possible to see this clear 'eye'. Sometimes it is even possible to see right down to the sea surface. Air descending in the very centre warms during its descent, giving rise to the name 'warm-core cyclone'.

Surrounding the eye is the 'eyewall' of highest convective clouds (giant cumulonimbus clouds) and the fastest winds. On animated images from satellites it is sometimes possible to see the way in which (in the north) the air is moving anticlockwise at the surface. The air rises within the eyewall and flows out of the top of the system. The outflow is clockwise in the northern hemisphere and again, animations of satellite images often show cirrus clouds flowing out from the centre.

Another factor in the formation of tropical cyclones is that there should be minimal vertical wind shear. (Vertical wind shear is a difference in wind speed at different heights. If the shear is too great, cyclones cannot form.)

Satellite images often show signs of the spiral bands of intense rainfall outside the eyewall. The rain from tropical cyclones, although it may be devastating, causing floods and landslides, is, in many regions, often essential to agriculture.

---

### Orographic

A term used to describe rain or conditions that arise when air is forced to rise over high ground. The increase of rainfall (orographic rain) is a common occurrence over the mountains of Wales and Scotland.

---

*Hurricane Isabel, photographed from the International Space Station on 15 September 2003. The cloud tops indicate the location of the spiral rainbands outside the eye.*

### Exosphere
The name sometimes applied to the upper region of the thermosphere above an altitude of 200–700 km.

# The Beaufort Scale

Wind strength is commonly given on the Beaufort scale. This was originally defined by Francis Beaufort (later Admiral Beaufort) for use at sea, but was subsequently modified for use on land. Meteorologists generally specify the speed of the wind in metres per second (m s⁻¹). For wind speeds at sea, details are usually given in knots. The equivalents in kph are shown for speeds over land.

**The Beaufort scale (for use at sea)**

| Force | Description | Sea state | Speed | |
| | | | Knots | m s⁻¹ |
|---|---|---|---|---|
| 0 | calm | like a mirror | <1 | 0.0–0.2 |
| 1 | light air | ripples, no foam | 1–3 | 0.3–1.5 |
| 2 | light breeze | small wavelets, smooth crests | 4–6 | 1.6–3.3 |
| 3 | gentle breeze | large wavelets, some crests break, a few white horses | 7–10 | 3.4–5.4 |
| 4 | moderate breeze | small waves, frequent white horses | 11–16 | 5.5–7.9 |
| 5 | fresh breeze | moderate, fairly long waves, many white horses, some spray | 17–21 | 8.0–10.7 |
| 6 | strong breeze | some large waves, extensive white foaming crests, some spray | 22–27 | 10.8–13.8 |

**The Beaufort scale (for use at sea)** – *continued*

| Force | Description | Sea state | Speed Knots | m s⁻¹ |
|---|---|---|---|---|
| 7 | near gale | sea heaping up, streaks of foam blowing in the wind | 28–33 | 13.9–17.1 |
| 8 | gale | fairly long and high waves, crests breaking into spindrift, foam in prominent streaks | 34–40 | 17.2–20.7 |
| 9 | strong gale | high waves, dense foam in wind, wave-crests topple and roll over, spray interferes with visibility | 41–47 | 20.8–24.4 |
| 10 | storm | very high waves with overhanging crests, dense blowing foam, sea appears white, heavy tumbling sea, poor visibility | 48–55 | 24.5–28.4 |
| 11 | violent storm | exceptionally high waves may hide small ships, sea covered in long, white patches of foam, waves blown into froth, poor visibility | 56–63 | 28.5–32.6 |
| 12 | hurricane | air filled with foam and spray, visibility extremely bad | 64 | 32.7 |

## The Beaufort scale (adapted for use on land)

| Force | Description | Events on land | Speed km h⁻¹ | m s⁻¹ |
|---|---|---|---|---|
| 0 | calm | smoke rises vertically | <1 | 0.0–0.21 |
| 1 | light air | direction of wind shown by smoke but not by wind vane | 1–5 | 0.3–1.5 |
| 2 | light breeze | wind felt on face, leaves rustle, wind vane turns to wind | 6–11 | 1.6–3.3 |
| 3 | gentle breeze | leaves and small twigs in motion, wind spreads small flags | 12–19 | 3.4–5.4 |
| 4 | moderate breeze | wind raises dust and loose paper, small branches move | 20–29 | 5.5–7.9 |
| 5 | fresh breeze | small leafy trees start to sway, wavelets with crests on inland waters | 30–39 | 8.0–10.7 |
| 6 | strong breeze | large branches in motion, whistling in telephone wires, difficult to use umbrellas | 40–50 | 10.8–13.8 |
| 7 | near gale | whole trees in motion, difficult to walk against wind | 51–61 | 13.9–17.1 |

**The Beaufort scale (adapted for use on land)** – *continued*

| Force | Description | Events on land | Speed km h⁻¹ | m s⁻¹ |
|-------|-------------|----------------|--------------|-------|
| 8 | gale | twigs break from trees, difficult to walk | 62–74 | 17.2–20.7 |
| 9 | strong gale | slight structural damage to buildings; chimney pots, tiles, and aerials removed | 75–87 | 20.8–24.4 |
| 10 | storm | trees uprooted, considerable damage to buildings | 88–101 | 24.5–28.4 |
| 11 | violent storm | widespread damage to all types of building | 102–117 | 28.5–32.6 |
| 12 | hurricane | widespread destruction, only specially constructed buildings survive | ≥118 | ≥ 32.7 |

# The TORRO Tornado Scale

The TORRO tornado intensity scale is based on an extension to the Beaufort scale of wind speeds. The winds speeds are actually calculated mathematically from the accepted Beaufort wind speeds. (Although the Beaufort scale was first proposed in 1805, it was expressed in terms of wind speed in 1921.) T0 corresponds to Beaufort Force 8, and T11 would correspond to Beaufort Force 30 (if such a force existed).

The TORRO scale is thus solely based on wind speeds, unlike the Fujita scale and the later, modified version, the Enhanced Fujita scale, which are based on an assessment of damage. In practice, wind-speed measurements are rarely available for tornadoes, and so, in effect, both scales are, perforce, based on an assessment of the intensity of damage.

| Scale | Wind speed (estimated) | | | Potential damage |
|-------|------|------|------|------------------|
| | mph | km h⁻¹ | m s⁻¹ | |
| F0 | 0–38 | 0–60 | 0–16 | **No damage.** *(Funnel cloud aloft, not a tornado)* No damage to structures, unless on tops of tallest towers, or to radiosondes, balloons and aircraft. No damage in the country, except possibly agitation to highest tree-tops and effect on birds and smoke. A whistling or rushing sound aloft may be noticed. |
| T0 | 39–54 | 61–86 | 17–24 | **Light damage.** Loose light litter raised from ground-level in spirals. Tents, marquees seriously disturbed; most exposed tiles, slates on roofs dislodged. Twigs snapped; trail visible through crops. |
| T1 | 55–72 | 87–115 | 25–32 | **Mild damage.** Deckchairs, small plants, heavy litter becomes airborne; minor damage to sheds. More serious dislodging of tiles, slates, chimney pots. Wooden fences flattened. Slight damage to hedges and trees. |

## The TORRO Tornado Scale (Continued)

| Scale | Wind speed (estimated) | | | Potential damage |
|-------|-----|-------|-------|-----------------|
| | mph | km h⁻¹ | m s⁻¹ | |
| T2 | 73–92 | 116–147 | 33–41 | **Moderate damage.** Heavy mobile homes displaced, light caravans blown over, garden sheds destroyed, garage roofs torn away destroyed, garage roofs torn away. Much damage to tiled roofs and chimney stacks. General damage to trees, some big branches twisted or snapped off, small trees uprooted. |
| T3 | 93–114 | 148–184 | 42–51 | **Strong damage.** Mobile homes overturned / badly damaged; light caravans destroyed; garages and weak outbuildings destroyed; house roof timbers considerably exposed. Some larger trees snapped or uprooted. |
| T4 | 115–136 | 185–220 | 52–61 | **Severe damage.** Motor cars levitated. Mobile homes airborne / destroyed; sheds airborne for considerable distances; entire roofs removed from some houses; roof timbers of stronger brick or stone houses completely exposed; gable ends torn away. Numerous trees uprooted or snapped. |
| T5 | 137–160 | 221–259 | 62–72 | **Intense damage.** Heavy motor vehicles levitated; more serious building damage than for T4, yet house walls usually remaining; the oldest, weakest buildings may collapse completely. |

**The TORRO Tornado Scale** *(Continued)*

| Scale | Wind speed (estimated) | | | Potential damage |
|-------|------|--------|--------|-----------------|
|       | mph  | km h$^{-1}$ | m s$^{-1}$ |              |
| T6 | 161–186 | 260–299 | 73–83 | **Moderately-devastating damage.** Strongly built houses lose entire roofs and perhaps also a wall; windows broken on skyscrapers, more of the less-strong buildings collapse. |
| T7 | 187–212 | 300–342 | 84–95 | **Strongly-devastating damage.** Wooden-frame houses wholly demolished; some walls of stone or brick houses beaten down or collapse; skyscrapers twisted; steel-framed warehouse-type constructions may buckle slightly. Locomotives thrown over. Noticeable debarking of trees by flying debris. |
| T8 | 213–240 | 343–385 | 96–107 | **Severely-devastating damage.** Motor cars hurled great distances. Wooden-framed houses and their contents dispersed over long distances; stone or brick houses irreparably damaged; skyscrapers badly twisted and may show a visible lean to one side; shallowly anchored high rises may be toppled; other steel-framed buildings buckled. |
| T9 | 241–269 | 386–432 | 108–120 | **Intensely-devastating damage.** Many steel-framed buildings badly damaged; skyscrapers toppled; locomotives or trains hurled some distances. Complete debarking of any standing tree-trunks. |

**The TORRO Tornado Scale** (*Continued*)

| Scale | Wind speed (estimated) | | | Potential damage |
|---|---|---|---|---|
| | mph | km h⁻¹ | m s⁻¹ | |
| T10 | 270–299 | 433–482 | 121–134 | **Super damage.** Entire frame houses and similar buildings lifted bodily or completely from foundations and carried a large distance to disintegrate. Steel-reinforced concrete buildings may be severely damaged or almost obliterated. |
| T11 | >300 | >483 | >135 | **Phenomenal damage.** Strong framed, well-built houses levelled off foundations and swept away. Steel-reinforced concrete structures are completely destroyed. Tall buildings collapse. Some cars, trucks and train carriages may be thrown approximately 1 mile (1.6 kilometres). |

# TORRO Hailstorm Intensity Scale

The Tornado and Storm Research Organisation (TORRO) has not only developed a scale for rating tornadoes (see pages 252–255) but also one to judge the severity of hailstorm incidents. This scale is given in the following table, but it must be borne in mind that the severity of any hailstorm will depend (among other factors) upon the size of individual hailstones, their numbers and also the speed at which the storm itself travels across country.

| Scale | Intensity | Hail size (mm) | Size comparison | Damage |
|-------|-----------|----------------|-----------------|--------|
| H0 | Hard hail | 5–9 | Pea | None |
| H1 | Potentially damaging | 10–15 | Mothball | Slight general damage to plants, crops |
| H2 | Significant | 16–20 | Marble, grape | Significant damage to fruit, crops, vegetation |
| H3 | Severe | 21–30 | Walnut | Severe damage to fruit and crops<br>Damage to glass and plastic structures<br>Paint and wood scored |
| H4 | Severe | 31–40 | Pigeon's egg > squash ball | Widespread damage to glass<br>Damage to vehicle bodywork |
| H5 | Destructive | 41–50 | Golf ball > Pullet's egg | Wholesale destruction of glass<br>Damage to tiled roofs<br>Significant risk of injuries |

**TORRO Hailstorm Intensity Scale** *(continued)*

| Scale | Intensity | Hail size (mm) | Size comparison | Damage |
|-------|-----------|----------------|-----------------|--------|
| H6 | Destructive | 51–60 | Hen's egg | Bodywork of grounded aircraft dented Brick walls pitted |
| H7 | Destructive | 61–75 | Tennis ball > cricket ball | Severe roof damage Risk of serious injuries |
| H8 | Destructive | 76–90 | Large orange > soft ball | Severe damage to aircraft bodywork |
| H9 | Super hailstorms | 91–100 | Grapefruit | Extensive structural damage |
| H10 | Super hailstorms | >100 | Melon | Extensive structural damage Risk of severe or fatal injuries to persons caught in the open |

# Twilight Diagrams

### Sunrise, sunset, twilight

For each individual month, we give details of sunrise and sunset times for the four capital cities of the various countries that make up the United Kingdom.

During the summer, especially at high latitudes, twilight may persist throughout the night and make it difficult to see the faintest stars. Beyond the Arctic and Antarctic Circles, of course, the Sun does not set for 24 hours at least once during the summer (and rise for 24 hours at least once during the winter). Even when the Sun does dip below the horizon at high latitudes, bright twilight persists throughout the night, so observing the fainter stars is impossible. Even in Britain this applies to northern Scotland, which is why we include a diagram for Lerwick in the Shetland Islands.

As mentioned earlier (page 9) there are three recognised stages of twilight: civil twilight, nautical twilight and astronomical twilight. Full darkness occurs only when the Sun is more than 18° below the horizon. During nautical twilight, only the very brightest stars are visible. During astronomical twilight, the faintest stars visible to the naked eye may be seen directly overhead, but are lost at lower altitudes. They become visible only once it is fully dark. The diagrams show the duration of twilight at the various locations. Of the locations shown, during the summer months there is astronomical twilight for a short time at Belfast, and this lasts longer during the summer at all of the other locations. To illustrate the way in which twilight occurs in the far south of Britain, we include a diagram showing twilight duration at St Mary's in the Scilly Isles. (A similar situation applies to the Channel Islands, which are also in the far south.) Once again, full darkness never occurs.

The diagrams show the times of New and Full Moon (black and white symbols, respectively). As may be seen, at most locations during the year roughly half of New and Full Moon phases may come during daylight. For this reason, the exact phase may be invisible in Britain, but be clearly seen elsewhere in the world. The exact times of the events are given in the diagrams for each individual month.

## Lerwick, Shetland Islands – Latitude 60.2°N – Longitude 1.1°W

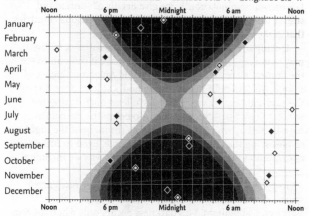

## Edinburgh, UK – Latitude 55.9°N – Longitude 3.2°W

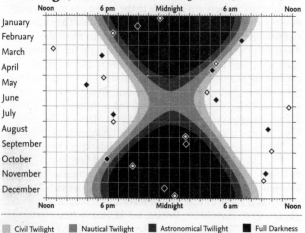

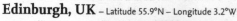

Civil Twilight   Nautical Twilight   Astronomical Twilight   Full Darkness

◇ Time of Full Moon   ◆ Time of New Moon

## Belfast, UK – Latitude 54.6°N – Longitude 5.8°W

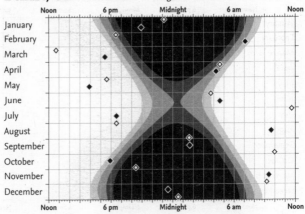

## Cardiff, UK – Latitude 51.5°N – Longitude 3.2°W

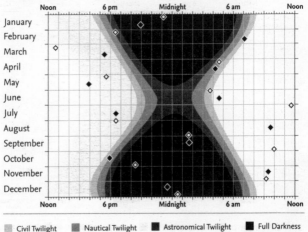

## London, UK – Latitude 51.5°N – Longitude 2.0°W

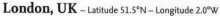
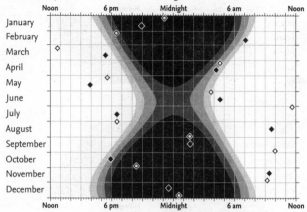

## St Mary's, Scilly Isles – Latitude 49.9°N – Longitude 6.4°W

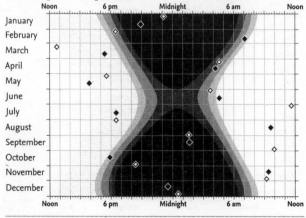

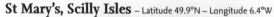

Civil Twilight    Nautical Twilight    Astronomical Twilight    Full Darkness

◇ Time of Full Moon    ◆ Time of New Moon

# Further Reading

**Books**

Chaboud, René, *How Weather Works* (Thames & Hudson, 1996)

Dunlop, Storm, *Clouds* (Haynes, 2018)

Dunlop, Storm, *Collins Gem Weather* (HarperCollins, 1999)

Dunlop, Storm, *Collins Nature Guide Weather* (HarperCollins, 2004)

Dunlop, Storm, *Come Rain or Shine* (Summersdale, 2011)

Dunlop, Storm, *Dictionary of Weather* (2nd edition, Oxford University Press, 2008)

Dunlop, Storm, *Guide to Weather Forecasting* (rev. printing, Philip's, 2013)

Dunlop, Storm, *How to Identify Weather* (HarperCollins, 2002)

Dunlop, Storm, *How to Read the Weather* (Pavilion, 2018)

Dunlop, Storm, *Weather* (Cassell Illustrated, 2006/2007)

Eden, Philip, *Weatherwise* (Macmillan, 1995)

File, Dick, *Weather Facts* (Oxford University Press, 1996)

Hamblyn, Richard & Meteorological Office, *The Cloud Book: How to Understand the Skies* (David & Charles, 2009)

Hamblyn, Richard & Meteorological Office, *Extraordinary Clouds* (David & Charles, 2009)

Kington, John, *Climate and Weather* (HarperCollins, 2010)

Ludlum, David, *Collins Wildlife Trust Guide Weather* (HarperCollins, 2001)

Meteorological Office, *Cloud Types for Observers* (HMSO, 1982)

Met Office, Factsheets 1–19 (pdfs downloadable from: http://www.metoffice.gov.uk/learning/library/publications/factsheets)

Watts, Alan, *Instant Weather Forecasting* (Adlard Coles Nautical, 2000)

Watts, Alan, *Instant Wind Forecasting* (Adlard Coles Nautical, 2001)

Watts, Alan, *The Weather Handbook* (3rd edn, Adlard Coles Nautical, 2014)

Whitaker, Richard, ed., *Weather: The Ultimate Guide to the Elements* (HarperCollins, 1996)

Williams, Jack, *The AMS Weather Book: The Ultimate Guide to America's Weather* (Univ. Chicago Press, 2009)

Woodward, A., & Penn, R., *The Wrong Kind of Snow* (Hodder & Stoughton, 2007)

## Internet links – Current weather

AccuWeather:       *http://www.accuweather.com/*
     UK:       *http://www.accuweather.com/ukie/index.asp?*

Australian Weather News:
     *http://www.australianweathernews.com/*

    UK station plots:
     *http://www.australianweathernews.com/sitepages/*
     *charts/611_United_Kingdom.shtml*

BBC Weather:       *http://www.bbc.co.uk/weather*

CNN Weather:       *http://www.cnn.com/WEATHER/index.html*

Intellicast:       *http://intellicast.com/*

ITV Weather:       *http://www.itv-weather.co.uk/*

Unisys Weather:       *http://weather.unisys.com/*

UK Met Office:       *http://www.metoffice.gov.uk*

    Forecasts:
     *http://www.metoffice.gov.uk/weather/uk/uk_forecast_*
     *weather.html*

    Hourly Weather Data:
     *http://www.metoffice.gov.uk/education/teachers/*
     *latest-weather-data-uk*

    Latest station plot:
     *http://www.metoffice.gov.uk/data/education/chart_latest.gif*

    Surface pressure charts:
     *http://www.metoffice.gov.uk/public/weather/surface-pressure/*

    Explanation of symbols on pressure charts:
     *http://www.metoffice.gov.uk/guide/weather/*
     *symbols#pressure-symbols*

    Synoptic & climate stations (interactive map):
     *http://www.metoffice.gov.uk/public/weather/climate-network/*
     *#?tab=climateNetwork*

    Weather on the Web:
     *http://wow.metoffice.gov.uk/*

The Weather Channel:
     *http://www.weather.com/twc/homepage.twc*

Weather Underground:
     *http://www.wunderground.com*

Wetterzentrale:       *http://www.wetterzentrale.de/pics/Rgbsyn.gif*

Wetter3 (German site with global information):
     *http://www.wetter3.de*

    UK Met Office chart archive:
     *http://www.wetter3.de/Archiv/archiv_ukmet.html*

## General information

Atmospheric Optics:
*http://www.atoptics.co.uk/*

Hurricane Zone Net:
*http://www.hurricanezone.net/*

National Climate Data Centre:
*http://www.ncdc.noaa.gov/*

Extremes:
*http://www.ncdc.noaa.gov/oa/climate/severeweather/extremes.html*

National Hurricane Center:
*http://www.nhc.noaa.gov/*

Reading University (Roger Brugge):
*http://www.met.reading.ac.uk/~brugge/index.html*

UK Weather Information:
*http://www.weather.org.uk/*

Unisys Hurricane Data:
*http://weather.unisys.com/hurricane/atlantic/index.html*

WorldClimate:
*http://www.worldclimate.com/*

---

## Meteorological Offices, Agencies and Organisations

Environment Canada:
*http://www.msc-smc.ec.gc.ca/*

European Centre for Medium-Range Weather Forecasting (ECMWF):
*http://www.ecmwf.int*

European Meteorological Satellite Organisation:
*http://www.eumetsat.int/website/home/index.html*

Intergovernmental Panel on Climate Change:
*http://www.ipcc.ch*

National Oceanic and Atmospheric Administration (NOAA):
*http://www.noaa.gov/*

National Weather Service (NWS):
*http://www.nws.noaa.gov/*

UK Meteorological Office:
*http://www.metoffice.gov.uk*

World Meteorological Organisation:
*http://www.wmo.int/pages/index_en.html*

---

## Satellite images

Eumetsat:
> http://www.eumetsat.de/

> Image library:
> http://www.eumetsat.int/website/home/Images/ImageLibrary/index.html

Group for Earth Observation (GEO):
> http://www.geo-web.org.uk/

---

## Societies

American Meteorological Society:
> http://www.ametsoc.org/AMS

Australian Meteorological and Oceanographic Society:
> http://www.amos.org.au

Canadian Meteorological and Oceanographic Society:
> http://www.cmos.ca/

Climatological Observers Link (COL):
> https://colweather.ssl-01.com/

European Meteorological Society:
> http://www.emetsoc.org/

Irish Meteorological Society:
> http://www.irishmetsociety.org

National Weather Association, USA:
> http://www.nwas.org/

New Zealand Meteorological Society:
> http://www.metsoc.org.nz/

Royal Meteorological Society:
> http://www.rmets.org

TORRO: Hurricanes and Storm Research Organisation:
> http://torro.org.uk

---

# Acknowledgements

| 29 | MV Braer | Document Scotland |
|---|---|---|
| 61 | WMO regions | WMO |
| 77 | Luke Howard etching | Wikimedia Commons |
| 96 | *Flaming June* | World History Archive / Alamy Stock Photo |
| 109 | *The Silent Highwayman* | Punch |
| 111 | Noctilucent clouds | Alan Tough |
| 131 | Robert Fitzroy | Liam White, Alamy Stock Photo |
| 143 | Upper Neuadd Reservoir | Wikimedia |
| 144 | First weather forecast | *The Times* |
| 145 | First weather chart | *The Times* |
| 160 | John Keats | IanDagnall Computing / Alamy Stock Photo |
| 165 | Edmond Halley | Granger/Shutterstock |
| 177 | Spoil tips, Aberfan | Wikimedia Commons |
| 209 | Honister Pass | Wikimedia Commons |
| 233–243 | Cloud types | Storm Dunlop |
| 247 | Hurricane Isabel | NASA |

# Index